Living
Among
Angels

In Memory of

Elwood Babbitt

Living Among Angels

*What would you do
if you were an angel?*

by David Zimmer and Luella Stroh

First Edition
Published in Great Britain
and the USA 2007
by
Tracker Press

ISBN 0-9546105-4-7

www.trackerpress.com

Cover design by David Foreman, Toast Design Consultancy
david@toastdesign.co.uk

Contents

Foreword

A word from the Editor

When Mother Nature was handing out skepticism, I was up there at the front of the queue. So when David approached me with the idea of publishing a book about angels actually living among us, I was not sure whether he should be taken seriously. But I was curious. If he was aware of these angels, what else did he know, and how could it benefit me?

Don't angels live on fluffy white clouds and watch over us with love and benevolence? Don't they have wings and halos and wear long white robes? How, then, could they be living among us?

Of course, this concept of angels challenged the logical part of my mind. Watching over us from atop fluffy white clouds? How silly! What about parts of the world where they rarely *have* clouds? And what about the wings idea? I suppose there's a certain amount of logic to that – after all, they have to get down here from Heaven somehow, right?

Perhaps David and Luella would be able to explain all of this in a way that makes sense. I decided to read their manuscript.

I found explanations that made sense to me. I also found a mine of other information that helped me to make sense of my own life: why I do the things I do, why things happen as they do.

Should you simply read this book and accept all that David and Luella say? No! Approach it with an open mind, and be willing to *consider the possibility* that what they say is accurate. Be willing to examine your own life and your actions in the context of their explanations. You might not agree with what they say. You might think it's rubbish. You might find it life-changing. You might simply not be ready for it yet. Whatever your reaction, I promise it will give you something to think about.

Pat Bensky

Preface:
Transcending Words

Everything contained in this book is based upon what Luella and I have personally experienced as opposed to what another has experienced. Rather than relying upon the contents of our memory to write it, we allowed ourselves to relive it and challenge our findings. Therefore, it is original to us. It will not be original to you until you allow yourself to venture into the unknown and experience it freshly. Few people realize that the majority of what is contained within their memory is not theirs because they haven't personally experienced it. It is what others have experienced, and most of it isn't accurate because it was obtained from others and interpreted by you. Not only that, but everything is constantly changing, including your perception. Unfortunately, the brain is unable to differentiate what you experienced from what another experienced. Similarly, the brain is unable to differentiate a real experience from an imagined one. Consequently, the brain treats everything contained within its memory as if it were real and accurate. The only way you can determine if it is still accurate and real is by allowing yourself to experience it freshly, uninfluenced by the contents of your mind.

An example of this phenomenon is when someone you like and respect frequently talks about a person you have never met. You may not be aware that the brain responds to that knowledge as if you had personally experienced it. Consequently, when you meet that individual for the first time, you respond to them based upon your friend's perception of them rather than seeing them freshly (uninfluenced by another) and responding accordingly. This problem is so prevalent that most people are unaware that many of their opinions and feelings about others were provided by somebody else.

Finally, you have unknowingly embellished or altered a lot of what you have and haven't experienced. These are the stories you have repeatedly told yourself as well as those you repeatedly told to others. They are so engraved in your memory that you perceive them as being real and accurate.

You can save yourself a lot of time, confusion, and emotional turmoil by embracing or playing around with the general concepts expressed in this book as opposed to allowing thought to fixate itself on some of the words and phrases used. Effectively communicating what we have experienced is extremely difficult due to the limitations of language and the different ways people process information.

The greatest hurdle in communicating what you have experienced occurs when you attempt to communicate anything about the non-physical (spirit) and the divine: that which is neither spirit nor physical. That is because you cannot rely upon physical images to communicate something that has no image, form, structure, or limitations. Therefore you have to use analogies and metaphors to describe that which is indescribable. Unfortunately, the brain has difficulty in recognizing and accurately interpreting analogies and metaphors. So it's easy for people to take them literally rather than figuratively and vice versa.

Furthermore, the message is limited by the messenger's vocabulary. Although we use words to communicate with others, the words we use are neither the described nor what was experienced. For example, the word "red" is a symbol that designates a particular color in the English language. But everyone has a different perception of what the color red is because everyone's perception of it is as unique as they are. Everything is relative to the eye of the beholder. A company named Pantone realized that and created a guide that allows people to precisely communicate a particular color on a particular surface.

Another hindrance in communicating is that some words in the English language have the same meaning but are spelled differently in various countries where English is the primary language. Besides that, some words in the English language do not mean the same thing in different countries or even in different areas of the same country. Furthermore, the meaning of any message is always altered when it is communicated from one person to another or translated from one language to another.

You are probably aware of other things that hinder people from effectively communicating with one another, especially in print to a greatly varied audience. So please pause and ask yourself, "How would I have written or expressed it?" whenever something in this book disturbs or frustrates you.

Acknowledgements

This book would never have come into existence if it hadn't been for Luella's involvement with the Akashic Records, and all the wonderful assistance provided by this marvelous soul.

Thank goodness Tracker Press had the wisdom and courage to publish it!

I am very thankful to my editor, Pat Bensky. In addition to editing the entire book, her valuable suggestions greatly assisted me in clarifying its contents. She's also responsible for the design and layout of the book.

I want to gratefully acknowledge David Foreman of Toast Design in the UK for designing the cover.

Although I am a key figure in the creation of this book, it was the life experiences of the angels featured in it that made it possible. Everything they experienced is just as important as what Luella and I wrote.

Finally I would like to express my gratitude to the thousands of souls I have encountered in life. It was through them that I was able to see and better understand myself.

David Zimmer

Absolutely nothing "is" what it appears to "be" when you use consciousness to question and challenge the illusions thought created.

Introduction

Geetings! I am David Zimmer. Luella Stroh and I see ourselves more as fellow students of life than as authors. *Living Among Angels* is a record of our involvement with angels living among us over the past twelve years. It is also a record of our findings and what we experienced examining creation, evolution and reincarnation for more than 25 years. Your investigations, experiences, and discoveries will be different to ours because everyone is unique. This is why we suggest that you view everything contained within Living Among Angels as a possibility.

I discovered the advantages of jotting down the contents of my mind so that I could see and examine what's contained there. This process freed my mind and allowed it to remain quiet, still, and attentive. Writing also keeps me focused and forces me to be more precise and accurate, which improves the integrity of my examinations.

This unusual book is also a multi-dimensional journey; a journey that simultaneously takes place on many levels. It is a journey that cannot be taken using thought, logic, or the analytical mind – it can only be taken using consciousness. Unfortunately, this concept can only be understood by consciously experiencing it. The limitations of thought cause thought to disagree with that statement and reject it because it is oblivious of its limitations. In other words, thought thinks and believes that it can understand it. Thus thought unknowingly and inadvertently prevents you from consciously experiencing this journey.

Luella and I decided to write *Living Among Angels* to gain a deeper and fuller understanding of what we experienced and discovered while assisting hundreds of angels. We hope it will stimulate your curiosity and cause you to examine your perception of angels – only you can determine if you're curious and adventurous enough to ignore thought and allow yourself to experience something new! The difference between personally experiencing the unknown and responding to what you know is illustrated in the following story about Galileo.

Galileo, the seventeenth-century astronomer, invited the best-known philosophers, theologians, and mathematicians of his day to experience what he observed through his telescope. He did this because he realized that they had to experience what he saw for themselves in order to realize the significance of his findings. Unfortunately, only a handful of great minds accepted his invitation. What devastated Galileo even more was how his guests reacted to the idea of examining the heavens through his telescope. Rather than simply looking into it, his guests debated for hours as to whether it was an ethical, moral, or logical thing to do. Unfortunately their energetic debating exhausted them, and they left without looking into the telescope. Many of those who never allowed themselves to experience what Galileo saw denounced Galileo's opinions on the motion of the Earth, judging them dangerous and heresy. A few of those who attended that meeting eventually developed enough curiosity to overcome their fear of doing something different, and what they subsequently experienced confirmed Galileo's findings.

One of Galileo's guests was Cardinal Roberto Bellarmino, a theologian who had recently become a "closet" astronomer. He quickly realized that he could not continue being an astronomer and practice his faith, because his findings from astronomy contradicted the doctrine of his faith. Upon examining the consequences of his dilemma, he concluded that his financial and political interests would be best served if he remained loyal to his faith, and so he resolved his dilemma by denouncing the truth and persecuting Galileo. As a key figure in the seventeenth-century Inquisition, he personally handed Galileo an admonition enjoining him to neither advocate nor teach Copernican astronomy, because it was contrary to the accepted understanding of the Holy Scriptures.

Like Galileo, Luella and I invite you to personally experience what happens when you embrace the contents of this book without judging any of it as being right or wrong. That can only be achieved by allowing yourself the opportunity to put aside everything you know about angels and approach it freshly, as if you'd never heard of an angel. In other words you are not relying upon or using the contents of your memory to observe it. Doing that will provide you with the motivation and resources needed to make your own investigation of angels. Your investigation will be different from ours or anyone else's. Every soul has its own path that cannot be traveled by another.

Living Among Angels is divided into three parts because it is a multidimensional journey. In many ways, *Living Among Angels* is like a road map or Global Positioning System (GPS): its purpose is to provide direction. It cannot participate or take the journey for the traveler – that's the traveler's responsibility.

The first part, *The Reality of Life*, is about the journey every soul makes to becoming an angel. You'll discover why it's almost impossible to recognize an angel and why most angels are unaware that they are angels. You'll also see why *you* could be an angel and not even know it!

The Angel Stories in Part Two are actual life experiences of incarnate angels. Their stories were obtained from the Akashic Records (see page 27). The Akashic Records (also known as The Book of Life) are an accurate, unembellished, and unbiased account of everything that has ever transpired in life or the universe. These amazing stories were provided by Luella Stroh, probably the best Akashic reader there is. That opinion is substantiated by the inability of anyone other than Luella Stroh to read veiled incarnations (see page 32).

You'll see how diverse and multifaceted angels are. Many of these life experiences are obscure or unknown, but others are well known, such as:

Nostradamus	Joan of Arc	Thomas Beckett
Betsy Ross	John Burgoyne	John Quincy Adams
Lafayette	Carausius	Marcus Aurelius

Their stories provide additional insight into various times in history such as Ancient Rome, the American Revolution, 16th-century England, and more. Many of the stories allow you to see how the Church, educators and historians have consistently altered and embellished history, and you'll probably find yourself questioning which version of history to believe: the version to be found in scholarly history books, or the one found in these pages. But that's the purpose of this book – to get you to think about, examine, and question the things you think you know!

Part Three, *Understanding What Life Is All About*, ties everything together and provides you with the tools we used. You will discover what has prevented humankind from being able to understand life. You'll realize from experience that the more you understand about angels, the more you understand about life and yourself.

Each section of *Living Among Angels* fulfills a particular purpose and has value; each is part of the whole experience. Why would you want to hinder or limit your investigation by ignoring any one of them?

Living Among Angels is the first of three books that examine humankind's perception of itself, the human experience, and the different kinds of souls incarnate on earth. *Living Among Aliens* is the second book in the trilogy.

It is a scientific and philosophical examination of aliens and humankind's perception of them. *Living Among The Trinity* is the third book. It is a self-evident examination of the Trinity and the Universal Laws of Life. Additional information about this amazing trilogy can be found on the Tracker Press website at www.trackerpress.com.

Finally, it is not our intention to insult your intelligence or deliberately step on your emotional toes. This trilogy was initially written for our benefit. This is why it's written in the style and manner in which Luella and I communicate with each other. It was the urging of hundreds of individuals that we have worked with that caused us to publish it.

The limitations of language and the physical limitations of our bodies may cause you some confusion and anxiety at times. For example, some of our definitions of words we use are probably different to yours. Consequently, we use these words differently than you do.

We expect you to be skeptical. It is always wise to treat anything that conflicts with what you think, know or believe as a *possibility* rather than a foregone conclusion. The former causes you to examine, question and challenge what you think, know and believe. On the other hand, the latter closes the door to discovery and prevents you from learning anything new. Regardless of the subject matter, there will always be differences in opinions, perceptions, and beliefs whenever you approach something from the known. Those differences cause thought to play the "I am right and you are wrong" game. Thought transforms a simple discussion into a contest because it is more interested in proving that it is right than examining and understanding anything that challenges it. The former creates conflict, whereas the latter creates harmony.

– David Zimmer

Part I

The Reality Of Life

Nothing in life is singular! Everything that "is" exists because of its relationship to something else.

Chapter 1

Are You An Angel?

It's amazing what you can discover about angels – or anything else for that matter – by merely changing your focus, altering your direction, and asking open-ended questions. Examining what you have in common with angels challenges your perception of them and allows you to see how much you are like them. So much so that you could be an angel and not even know it! That's a difficult concept to grasp if you see yourself as being different from angels. It may surprise you to know that the majority of angels see themselves as being different from angels as well. Prior to our working with the angels in Part II, none of them were aware of their status and true identity. You are unaware of your true identity if it is based upon your physical body, your occupation, your beliefs, and other human characteristics. In other words, your perception of yourself constitutes your identity, whether you are aware of it or not. This can be demonstrated by:

- How you describe yourself

- How you interact with life (your behavior)

- What you defend or become defensive about

Your perception consists of everything you know, believe, and experienced that you have not questioned, challenged, and thoroughly examined. Your perception determines:

1) What you see and how you see it (interpret it)

2) What you experience and how you interpret it

3) And how you respond to one and two

Our experience in working with angels has allowed us to see that understanding yourself and realizing your true identity is essential to knowing what angels actually are, and understanding their function or purpose. And the key to understanding yourself lies in your willingness to use consciousness to challenge and thoroughly examine your perception. It also substantiates the notion that you can change what you experience and improve the quality of your life by changing your perception.

This part of the book presents a **possible alternative** to how you perceive yourself and your relationship with life. It will enable you to change your perception if you allow yourself to embrace it and attentively observe how it affects you and how you respond to it. Some parts of it will disturb you more than others. You'll probably experience more emotional discomfort such as apprehension, agitation, and anger the more thought defends its perception. Thought is the language of perception. The less you examine what you know, believe and what you have experienced, the more problems you experience. Thought defends everything it knows, even if it is false or inaccurate, because it identifies itself as being that knowledge. For this reason it's important to realize that everything that is real or accurate can withstand any challenge or test. That which is not accurate will change or cease to exist.

Here's another way of looking at all this. The human body physically sustains itself by processing air, food, and water. It discharges the waste – the by-product of this process. In fact, the health and wellbeing of the body is dependent upon what it does and does not consume, process, and discharge. The right and freedom to choose (free will) make you responsible for everything you physically consume, and how you view that responsibility determines how wise you are about what you consume. Consciousness and the body are responsible for processing and discharging what you physically consume. This is why you might not give any thought to how or whether the body is processing and discarding what you consume until the body malfunctions or is ill. The Universal Law of Action and Reaction causes you to experience favorable consequences when you fulfill that responsibility and unfavorable consequences when you do not. It's an example of the indestructible relationship associated with rights, responsibilities, and consequences.

Although you probably realize that, you may not be aware that the health and wellbeing of the body, what you experience, and the quality of your life are dependant upon what you *mentally* consume, process, and discharge. Here again, the responsibility and consequences associated with free will make you completely responsible for everything you mentally consume. And it is the mind and consciousness' responsibility to process and discharge what you mentally consume. All this becomes crucial when you realize that your perception is everything you don't question, challenge, and thoroughly examine before you mentally consume it. And it becomes even more critical when you realize that you do not have the foggiest clue as to how often you did this and how much mental junk food you consumed. Fortunately your vocabulary, behavior, everything you experience, and the quality of your life indicate the amount and type of mental junk food you have consumed since you started learning a language. Mental junk food causes the body to be physically and mentally ill and it can even bring about its demise.

You can choose to continue doing what you have been doing and experience the consequences. Or you can choose to start discovering and eliminating the mental junk food that you have unknowingly consumed.

The following questions may help you decide which to choose.

- Is the quality of your life improving from month to month, year after year, or are you doing everything you can just to stay afloat?

- Are you accomplishing what you want to accomplish, or does it appear that what you want always seems to be just out of your reach?

- Is there order and peace in your life or is it chaotic, experiencing one crisis after another?

- Other than what you have personally experienced, what do you really know for sure?

- Would you bet your life on the validity and accuracy of what you have mentally consumed? Whether you realize it or not, you are doing that when you assume that what you have consumed is valid and accurate.

Creation

Editor's note: You may find the language in this section rather archaic. That's because it is drawn directly from the Akashic Records by Luella Stroh.

In the beginning before the beginning of no beginning, there was "naught". It was absence absolute; void of those non-physical and physical elements that would evolve to stimulate, propagate, and nourish life as it is known today. Even their conception was yet to be.

It was perfection absolute – for in complete absence there exists not imperfection. That was yet to be, arising out of and created by the discovery of the fallibility of man while experiencing the physical world.

In this state of absolute absence and absolute perfection everything that is recognized and accepted as contributors to life of this present moment was nonexistent. And yet within that absolute state of perfection, and unaware of its existence, was the Atom Absolute. Do not waste one precious moment on idle speculation of how the awakening of this creative force came to be, for none were present to bear witness. Imaginations of the mortal mind cannot possibly recreate nor even begin to fathom the majesty of that event, for in this sanctity the Atom Absolute aroused as from an ancient sleep and realized it existed. Within this realization was born a need to express its momentous discovery. At first, an imperceptible stirring as the need kept growing and expanding until it filled this forbearer of all that was yet to be. Eventually the Atom Absolute began to pulsate. A steady rhythm now emitted from its being in this womb of complete perfection and absolute absence. Its pulsations registered as shock waves and each one, unrestrained, met no resistance. They crashed into one another, stretching over expanses beyond comprehension, only to rapidly and randomly fold back over one another to conjoin and magnify their force.

It was now chaos absolute. Pummeled, buffeted, and unable to resist the violence set into motion by is own creation, the Atom Absolute now pulsated even more rapidly. The intensity of the storm increased. It was as though this former sanctuary was attempting to expunge this unknown intruder from its presence so that it could return once again to its state of quiet and peace, to regain the absolute absence and perfection of its original design.

This was not to happen easily. One more intrusion even greater than the pulsations now emanated as the shock waves intensified by joining one with another and reverberating back, crashing into their creator. A primordial groan was elicited as it played out upon those resounding forces; this sound was as though wrenched from its bowels. Tortured and warped – the sound of matter being rendered asunder. "Yaaaa, rrrrr, wh wh wh wheeee."

An explosion was inevitable and unavoidable. Unable to contain the need to express, filled completely with this intense desire and no longer able to withstand the reaction of its own creation, the Atom Absolute shattered, sub-dividing itself into three – each now as complete and perfect as the original, each infused with the need for expression, and each annealed by the horrific force of its creation.

As an awakening consciousness that had but a moment before it discovered it existed, so also contained within each was that awareness. Thus, each pulsated in accordance with that vibratory awareness, and the resulting shock waves continued, unrelenting. The three were mercilessly carried up and hurled about, only to encounter yet another wave of energy which would again fold in upon them or toss them into the throes of oncoming energy.

Unable to exert any control over the storm raging without, each of the three instead initiated an internal examination, waging a struggle to understand what it meant to exist. Each questioned the purpose of existence and a means whereby it could express the absolute joy of that discovery. And within each was registered the pulsations of the others, continuing until the dawning of revelation that each was not alone. From somewhere pulsations unsynchronized with those of the examiner were as though crying out to each other. As each was confronted with this emerging awareness and attempted to understand the accompanying implications of this reality, their pulsations slowly attuned to each other until a rhythm was born. From the synchronization of the pulsations, the three came into an alignment, thereby establishing a rotation, each still seeking to come to know itself as well as one another.

NOTE: The phrase "Atom Absolute" does not pertain to a physical atom or sub-atomic particle. The Atom Absolute is the personality of God and the origin of every personality. It is the expression of the Consciousness Absolute, the consciousness of God. The division of the original Atom Absolute into three Atoms Absolute (Will, Love, and

Intelligence) created balance and ensured that life would always be balanced.

It is also important to realize that words are never the described! They merely represent what "is". For example, the word "naught" is merely an image that represents a plane of existence that cannot be imagined or comprehended by thought. Only that which is part of "naught" is able to experience and understand something that is neither physical nor non-physical. Likewise, only consciousness is able to understand the non-physical or spirit world because it is part of that world, as well as the physical world. On the other hand, thought is limited to the physical world.

Not all words apply to the physical world. For example, there is a vast difference between "awareness" and "conscious awareness". The word "awareness" means being aware of what stimulates and registers upon the body's physical senses. "Conscious awareness" is being simultaneously aware of what stimulates and registers upon mankind's physical, mental, and spiritual nature. The mind is perfectly quiet, still, and alert when you are consciously aware. There is no "self," "me," or what you identify yourself as being. Thus, the contents of your memory are not coloring or influencing what you see or observe.

Evolution

Nothing in life is singular! Everything that "is" exists because of its relationship to something else.

That statement means that absolutely nothing can exist by itself. Even before the beginning of no beginning there were three things: "naught", the Atom Absolute and the Consciousness Absolute. Similarly, hot cannot exist without cold, and up exists because down exists. This principle also means that manifestation involves consciousness and the personality. And everything that is manifested simultaneously creates two things. The desire to express and the desire to understand are mirrored expressions of each other; each is dependent upon the other for its existence. Mirroring is important because it keeps things balanced. Life is balanced because the trinity and everything it creates are mirrored and balanced. And balance prevents anything singular (unbalanced) from existing.

You need something other than yourself to serve as a mirror to see and understand yourself. You can also understand yourself by experiencing what you are *not*. These two statements appear to be different because they are mirrored. You can easily observe how something mirrored appears to be different by holding up printed text in front of a mirror: the mirrored image of the text is hard to read because it is backwards, yet the text is the same. Later on you will see how every human being serves as a mirror, thereby assisting you in seeing and understanding yourself.

The non-physical world (spirit) is a mirrored expression of "naught" and the physical world is a mirrored expression of spirit. These mirrored expressions allow the three atoms absolute to understand themselves by being and experiencing what they are not in the spirit and physical worlds.

The Personality And Permanent Atoms

The word "soul" is an ambiguous term because everyone has a different image, concept, or perception of it. Either all are accurate or none of them are accurate. That is because life treats everything in its domain exactly the same. It does not have any favorites or grant special favors.

Contrary to popular belief, the soul is the consciousness of humankind, just as the Consciousness Absolute is the consciousness of the trinity. There is only **one soul** and it is the mirrored expression of the Consciousness

Absolute. The soul is shared by all of humankind just as the trinity shares the Consciousness Absolute.

Just as the three Atoms Absolute (the trinity) are the expression of the Consciousness Absolute, countless billions of permanent atoms (personalities) are the expression of the soul. The word "countless" is used because it is impossible for the mind to comprehend them all. Your personality is your real identity because it is as unique as DNA. Unlike behavior, your personality is indestructible and remains the same from one incarnation to another and whether it is incarnate or not. Thus, your real identity is divine and possesses the same potential as the Atoms Absolute. Are you beginning to see why it has been stated that "Man is created in God's image"?

The only way the personality can be described is by stating what it is not because it is divine, meaning neither physical nor non-physical. Unlike the human body, it is neither male nor female. It's not affected by time, space, or gravity. It has no image, density, or limitations. The absence of limitations and restrictions allows the personality to do numerous things that are considered to be impossible by science, intellect, or thought. For example, it can simultaneously be in the past, present and future, as well as in different locations.

For clarity and the sake of this discourse, the terms "individual personalities" and "permanent atoms" are synonymous. All personalities are individual personalities except the three Atoms Absolute (Will, Love, and Intelligence). *Resident personalities* are new to incarnating and are serving their first tour of duty.

Like the Atoms Absolute, your personality desires to express and understand itself, and the only way it can obtain that understanding is by experiencing what it is not in the non-physical and physical worlds. Spirit is the three-dimensional vehicle the personality uses to experience the non-physical world. It is a mental reality created by the personality. The non-physical world is a non-progressive world because it cannot be physically experienced. It can only be experienced mentally, abstractly, and emotionally. It is the world where the personality reviews its last physical experience and plans its next incarnation.

The physical world is an evolutionary reality or progressive world because the personality's vehicle (the human body) can physically experience it. The body's senses allow the personality to experience everything its vehicle experiences. In fact, the personality is confined to the same limitations as its vehicle. In order for the personality to "be" what it is not so that it can understand itself it has to "perceive" that it is something other than what it is. Incarnating veils the personality's identity from itself, confines it to the limitations of its vehicle and what it experiences, and subjects it to thought's influence. All these cause the personality to falsely identify itself as being its vehicle. In other words, an incarnating personality's vehicle serves as a disguise that conceals the personality's identity from others and itself. When it is time for the personality to awaken from the illusion it created by incarnating, life lifts the veil and the personality realizes the illusion and thereby frees itself from it.

The relationship between the personality and its physical vehicle is similar to the relationship you have with your automobile or truck. Unlike thought, the personality isn't always concerned about its vehicle's comfort. Like you, the personality selects its vehicle based upon its ability to fulfill its objective by providing what it desires to experience. The vehicle gets replaced when it ceases to fulfill the personality's purpose.

The personality's ability to replace its vehicle allows it to experience what it needs in order that it might obtain the understanding it desires to experience. Obtaining it requires numerous vehicles because its vehicle limits what it can experience. For example, understanding "compassion" involves experiencing:

1) being compassionate

2) being void of compassion

3) receiving compassion

4) receiving the absence of compassion.

Thought is unable to see and understand that the personality cannot experience all four perspectives and understand it in a single incarnation. Everything the personality has experienced prior to serving as an angel has allowed it to understand that – this is why the personality is willing to devote over 21 incarnations to experiencing and understanding a single concept or field of service while serving as an angel.

Relationships

Life is the whole and everything that comprises the whole. It is comprised of relationships within relationships. Each of these relationships is an evolving process. Each process is a system within a system. All of these relationships, processes, and systems mutually share rights and responsibilities that allow them to develop their potential and fulfill their purpose. This is another way of saying, "Nothing in life is singular; everything exists because of its relationship to something else".

The relationship between the Atoms Absolute and Consciousness Absolute is identical to the relationship between the permanent atoms and the soul. Both of these relationships are balanced and indestructible. The latter has a relationship with the former because it is a mirrored expression of the former. Here again, what applies to one applies to all. Balanced relationships keep everything balanced and prevent anything from happening by chance, accident, or coincidence.

The personality exists because of consciousness, and the consciousness exists because of the personality. The personality is responsible for using the resources of consciousness to develop and express its unlimited potential. Likewise, consciousness is responsible for using the expressions of the personality to develop and express its unlimited potential. Both the personality and consciousness are accountable for everything they do because they have complete control over themselves. Consciousness, the personality, and their relationship are autonomous. That means life is self-governing; no entity directs or is in charge of life. Universal laws that are fair, impartial, and just govern life. These laws are absolute in that it is impossible for anything to alter, break, or circumvent them. They also make it impossible for life to pick and choose what it will offer and to whom.

Rather than move straight on to the next paragraph, you might want to consider pausing and carefully pondering the significance of all this. What does it all mean to you? Does it mean that you have absolute control over your life in respect to what you experience, how it affects you, and the quality of your life? If it does, what facilitates and what hinders you from improving the quality of your life? Or does it mean something else? If so, what, and how are you affected by it?

Resistance

The purpose of life and the personality is to develop mutually and express their unlimited potential. Resistance develops potential and gives it worth: obtaining or achieving anything of value involves overcoming some form of resistance through the expenditure of time, resources, and energy. The greater the potential worth, the greater the resistance encountered.

Resistance can be physical, mental, emotional, or a combination of all three. The limitations of the physical body create resistance, especially those commonly seen as a handicap. Your environment provides the physical resistance needed to develop muscle and stamina in the human body. Thought, perception, and illusion are life's formidable adversaries: they provide the mental and emotional resistance needed to develop values, character, and the quality of your mind. You'll realize that fact when you observe, question, examine, and challenge the contents of your mind.

But how do you do that? Unfortunately, you can't look in and physically examine the contents of your mind as if it were a drawer or filing cabinet. Instead, observe how everything you experience affects you physically, mentally and emotionally. It is also important to realize that the mind will protect itself from experiencing additional emotional pain by denying or concealing emotionally disturbing memories or experiences.

New At Incarnating

The vast and greatly diversified spectrum of consciousness is expressed by every personality that is incarnate. Each is as unique and distinctive as a fingerprint, which means that everything in life and its expression is original and never duplicated. It also means that nothing in life is ever wasted. Personalities who appear to have the same abilities differ in how they express or develop their potential. They also vary greatly in how often they incarnate or, indeed, if they have ever been incarnate. You'll probably be surprised to learn that some personalities have never incarnated and see no reason for doing so. How experienced they are and their level of conscious awareness determines the frequency with which personalities incarnate. Those that have obtained a higher level of conscious awareness and are more experienced incarnate more frequently than others.

All personalities incarnating for the first time are attracted to a planet that will best serve their needs and interests. Depending upon the personality's level of consciousness and ability to develop its potential, its first tour of duty or assignment on that planet will involve more than 21 incarnations. The tour of duty ends when they have destroyed the illusion they initially created when they incarnated. These personalities are called "resident personalities" so that they can be distinguished from other personalities. Throughout their tour of duty in the physical world resident personalities receive the assistance of older and more experience personalities, who serve as guides or mentors. That's important because newly incarnating personalities lack experience, just as young children lack the experience acquired by their parents and grandparents. Thus, they are less aware of potential problems and the possible consequences of their actions.

Prior to incarnating for the first time, everything the personality knows is "inexperienced knowledge" or academic knowledge. The experience of incarnating allows the personality to realize the difference between "inexperienced knowledge" and "experienced knowledge." The former is mental in that it has not been experienced, whereas the latter has been gained from experience. Experience challenges academic knowledge, elevates it, and transforms it into experienced knowledge. In turn, experience and consciousness challenge experienced knowledge, elevate it, and transform it into understanding. Likewise, experience and consciousness challenge what is understood, elevate it, and transform it into wisdom. Wisdom is the wise and effective use of understanding. It is the personality's objective because wisdom allows the personality to understand itself, life, and its relationship with life.

Prior to physically incarnating, the personality examines every possibility that it will probably encounter when it is incarnate – the process is similar to creating an interactive video game that has practically unlimited possibilities. Misconceptions are inevitable because even the wisest and most experienced personality is unaware of its misconceptions or – a single misconception can cause the personality to experience hundreds of unseen problems. Additional problems occur because it is impossible for the personality to anticipate destiny's (life's) involvement. Inaccuracies, misconceptions, and what are falsely seen as mistakes provide the personality countless opportunities to correct its perception and to gain additional understanding. Upon completing its examination, the personality is attracted to a vehicle in an environment that will allow it to experience what it desires to understand. Ultimately, karma and the Law of Attraction determine every incarnation. The personality then works out various agreements and arrangements with every personality that will mutually participate and share in that life experience of its vehicle. So it isn't by chance, accident or coincidence where you are born, how you are raised, or the environment you are exposed to.

The incarnation process is similar to experiencing amnesia in that the personality is unable to access its memory. The absence of memory allows the personality to spontaneously respond to its environment and everything the personality chooses to experience. The absence of memory also means that the personality has no identity because its memory is its identity. Not having an identity causes the personality to falsely identify itself as being its vehicle, thereby completing the process of being what it is not so that it can understand itself.

All the knowledge the personality has doesn't prepare it or help it realize what it will experience when it's incarnate because it has never experienced physical sensations or limitations. Upon completing the transformation from divine to physical, the personality wakes up in a strange and limited environment. It doesn't understand the sensations and limitations it is experiencing or why it is experiencing them. Overwhelmed and fearful of what it is experiencing, it is very easy for the personality to become traumatized. Prior to that happening, life withdraws the personality from its vehicle, thereby terminating the pregnancy. We call this tragic event a "miscarriage".

Because you probably don't understand this process, you don't realize that life saved the personality from being severely traumatized. Several incarnations are required for the personality to overcome its anxiety about incarnating, so life will cause the personality to withdraw from its vehicle several times before it becomes comfortable with the "human experience". Depending upon the personality, it may happen shortly after the personality experiences the birthing process during its second incarnation. This event is commonly called sudden infant death syndrome (SIDS). During the third incarnation, life (destiny) may cause an apparent accident or illness that ends the life of the personality's vehicle prior to reaching adulthood. Each subsequent incarnation involves the personality experiencing more of the human experience for a longer period of time.

Maturing Personalities

The personality reviews each incarnation after it has withdrawn from its vehicle. It compares what it actually experienced in a progressive world against what it mentally experienced in the non-progressive world, and the differences observed between them assist the personality in planning its next incarnation. Rather than assume that its perception is accurate, most personalities will compare their findings against the Akashic Records. That's important because the personality's review is so thorough that all knowledge of what previously transpired is completely consumed – as depicted by the Phoenix. Consequently, the personality has no knowledge of what previously transpired when it reincarnates.

There are instances when a personality may receive a few brief glimpses from previous incarnations, just as they may have an occasional deja vu experience. The source of this information and the manner in which it is obtained is different from recalling something from memory. This also applies to sensitive children prior to when they are dependant on a language to communicate, or become influenced by thought.

Each life experience assists the personality in seeing that it is going to unknowingly and unintentionally create some misconceptions. In turn, those misconceptions are going to cause the personality hundreds of problems in various ways. The maturing personality uses that understanding to plan each subsequent incarnation more wisely, which raises the personality's level of conscious awareness.

Judging what another individual does or experiences may prevent you from seeing that it is often necessary for its personality to have them experience what it desires to understand. For example, a mature personality may plan to have its vehicle become an alcoholic so that it can understand addiction. Likewise, the personality may deliberately incarnate in an abusive family and have its vehicle become abusive to better understand anger and control. Depending upon what the personality desires to experience, it may take several incarnations of accumulated misconceptions for it to experience what it desires to understand.

Eventually the personality matures and raises its conscious awareness enough to start questioning and challenging its perception and what it thinks it knows. That process allows the personality to realize that its perception is flawed. In turn, it realizes its real identity and destroys the illusion it created.

Unlike the personality, the Consciousness Absolute is aware of all the personality's misconceptions and the problems they create. Those errors disrupt the balance of life and cause life to restore it by making the personality aware of all its misconceptions while it is incarnate. This process is called a "review incarnation."

A review incarnation is radically different from other incarnations because the personality is being confronted with all its accumulated misconceptions while it is incarnate. The process is equivalent to experiencing numerous life experiences in one incarnation. The nature, intensity, and number of these examinations cause the vehicle to quickly move from one career or job to another, from one location to another and from one interest to another. All this movement destroys many of the personality's attachments and prevents it from establishing new ones.

On a scale of 1 to 1,000, the average conscious awareness of a resident personality is less than 200. From one incarnation to another, the average increase in a resident personality's conscious awareness is 2 points. During a review incarnation, a resident personality will increase its level of consciousness awareness by about 20 points. That is about 10 times the amount of change that an average person experiences in a lifetime. On average a resident personality will increase its conscious awareness by 50 points by the end of its tour of duty. Personalities serving a tour of duty as an alien or an angel have an advantage over a resident personality in that they have experienced and effectively dealt with the illusion of the non-physical and physical worlds. Thus, the personality of an alien or an angel will raise its level of conscious awareness by about 100 points by the end of their tour of duty.

Initially it is difficult for any individual participating in a review incarnation to understand why they are so different from most people. Most personalities experiencing a review incarnation see themselves as being outsiders, odd, and don't feel that they're part of anything. Many of them try to escape their situation through alcohol, drugs, gambling, or some other vice. In turn, society views them as being inferior and criticizes them. Society sees them as being indecisive, unstable, and unreliable. Most of those criticizing these individuals do not realize that these personalities are actually functioning on a higher level of conscious awareness than them.

What the personality experiences during a review incarnation allows it to raise its consciousness so that it can free itself from the "illusion of life" and realize its true identity. The personality is released from its attachment to the planet or assignment.

What was once an inexperienced and naive personality is now a more experienced and mature personality functioning at a higher level of conscious awareness. It is then given an opportunity to assist younger and less experienced personalities on another planet. Upon accepting its new assignment it becomes an alien personality because it is now an alien or foreigner to that planet. Depending upon the personality, an alien tour of duty will consist of about 24 physical incarnations that will raise its level of conscious awareness by about 100 points.

For clarity and the sake of this discourse, the terms "aliens" and "alien personalities" pertain to individual personalities that are more experienced than resident personalities and who are serving a tour of duty on a foreign planet – one that is different from their first tour of duty.

As before, the incarnation process causes the personality to lose its memory. Thus, the alien personality is unaware of its identity or what it previously experienced when it begins its assignment. The absence of memory allows the personality to falsely identify itself as being its vehicle and fulfill its assignment. Being in the "unknown" allows the alien personality to unknowingly assist others and gain additional experience as it works to free itself from the illusion of life. As when it was a resident personality, it completes its tour of duty when it destroys the illusion it created by incarnating, and realizes its real identity. In turn, it is provided with new opportunities to serve and raise its level of consciousness. One of them is being an angel.

The angels incarnate on Earth are like mankind; they come in all shapes, sizes, and colors. Just as you are unaware of them, the vast majority of angels are unaware that they are angels. Therefore you could be an angel and not even know it – only those angels participating in a review incarnation are consciously aware that they are angels. Upon completing their review incarnation they move on to a new adventure and are replaced by newly assigned angels.

Depending upon the personality's nature, interests, and level of conscious awareness, there are hundreds of different ways a personality can serve. Being an angel is just one of them. The range of different ways a personality can serve is similar to the wide range of occupations available in the physical world based upon your personality, interests, and level of education. Thought, conditioning, and perception can cause you to view some occupations as being more valuable, interesting, or important than others. What you may fail to see is that they all have equal value and importance because they are all unique.

Guardian Angels

The belief in guardian angels can be traced back through antiquity. Unfortunately, few people realize that there are numerous misconceptions associated with any belief. For example, the dictionary defines an angel as being a divine being who acts as a messenger of God. That is a very poor definition because every personality is divine and every personality is a messenger of God.

The majority of mankind mistakenly believes that those on the "other side" or in spirit are wiser and more intelligent then they are. They believe that because they are not aware that the majority of individuals who pass over to the other side merely pick up where they left off in the physical. The act of losing the physical vehicle and gaining a spirit vehicle does not make you wiser or more intelligent: it merely frees you from the limitations of the physical vehicle.

Many personalities that are perceived to be guardian angels are not. Some of these personalities have no practical experience because they have never been incarnated. They are unable to see why any personality would incarnate and experience all the emotional pain, sorrow, and conflict associated with incarnating when they don't have to. When it comes to assistance, wouldn't you prefer an experienced personality to an inexperienced one?

There are numerous references to angels and guardian angels in both the Old and the New Testament of the Bible. But, men wrote the Bible. The personalities that dictated what to write wanted mankind to realize that he is never "alone" and that there are other personalities assisting him, even if he is unaware of them or cannot see them. Therefore, they provided those transcribing the Bible with images designed to assist humankind in distinguishing these helpers from mortals. These images gave birth to the belief that angels have wings and halos.

The personality is never without "spiritual guidance" either in the physical or non-physical world. Jesus pointed that out when he said, "Man is not meant to live alone." Aliens and angels who are not incarnate help humankind when he is incarnate. A better word for these personalities would be "mentor" rather than guardian angel, because not all of these helpers are angels. Sometimes humankind receives help from those who are even more consciously aware than angels.

What would happen if you viewed non-incarnate helpers the same as those who assist you in the physical world? Those who assist you as an adult are commonly called mentors, personal trainers, coaches, and trainers. And because you don't call them guardian angels, you see and react to them differently. Similarly, when you were a child, your closest mentors and guardians were your parents and relatives. If you pay close attention to how you feel about each of these labels you will see that you react to them differently because you perceive them to be different.

Based upon experience I have always found it more beneficial and rewarding to deal with the "reality of life" or "what is" as opposed to the "illusion of life".

Personality Groups

Bonds are formed between personalities who have worked closely with each other in previous incarnations. Individual personalities who share a mutual bond are called "companion personalities" and belong to various personality groups. Companion personalities incarnate together from one incarnation to another. The most common personality group is comprised of relatives, friends and business associates.

The family unit offers the greatest potential for growth and understanding. Labels, social pressures, and your perception of "family" motivate and greatly assist all personalities within the family to work out their differences and understand each other. Family members will also switch roles with each other from one incarnation to another to better understand what each experienced. A change as simple as being the only child, the oldest child, middle child, or the youngest child alters what the personality experiences.

In many companies, the majority of employees have worked with each other in several incarnations. Regardless of how qualified you may be, it's difficult to get a job with these companies if you are an outsider. The residents of a community, state, and country comprise other types of personality groups. Some communities are comprised entirely of companion personalities who have lived together for several incarnations. The residents of these communities may unknowingly shun or make life difficult for outsiders to reside within their community. The older the personality group, the more likely it will change religions, races, and nationalities. Most companion personalities belong to several different groups.

That unexplainable feeling you get when you meet someone for the first time is a good indication that you knew them in a previous incarnation. If you immediately like, trust, and feel comfortable around them you're responding to previous favorable experiences with them. If you immediately dislike, don't trust, or feel uncomfortable around them you're responding to previous unfavorable experiences with them.

Chapter Two: Other Concepts

Karma

There are just as many misconceptions about karma as there are about the personality. The Law of Cause and Effect is the law of balance. Every misconception the personality has causes the personality to unknowingly oppose the Universal Laws that govern all life. That opposition causes the personality to be imbalanced and disrupts life's balance. Karma is the process life uses to restore the personality's imbalance. Life responds to what caused the imbalance by creating events, situations, or conditions that assist the personality in realizing its misconception. This realization is important for it allows the personality to correct its perception, eliminate the conflict, and regain its balance.

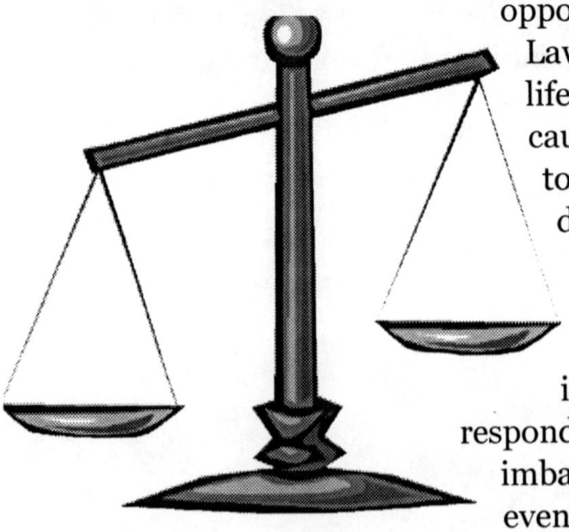

Your perception defines your reality and determines how you respond to it. Your response or reaction to your reality is called "behavior" because it is a conditioned response. The flaws in your perception alter or transform what is into what it is not; they create a difference between what is and how it is perceived. The greater this difference is, the more disruptive your behavior. In biblical times these errors were called sins. And because everyone errs, everyone was considered a sinner. Unfortunately, today "sinner" has a different connotation.

The personality's perception defines what its vehicle (the human body) will experience. Like its vehicle, the personality has misconceptions or flaws in its perception. Karma deals with the source of the error, as opposed to its expression or reflection. It is an impersonal response to the personality's flawed perception. Therefore, it never creates a contractual obligation. Karma is neither bad nor a punishment. It guides the personality through life and keeps it on course just as feedback guides and keeps a missile on target. The beauty of karma is that it treats every personality the same. What applies to one, applies to all. No personality is exempt or privileged.

Karma pertains to the personality's perception, whereas behavior pertains to the perception of the physical body. Focusing your attention on your actions or behavior may cause you to falsely associate karma with behavior and prevent you from seeing the source of your behavior. If karma was based upon behavior, it would create additional problems and conflict. For example, the late Senator Joseph McCarthy's "witch hunt" over communism ruined the reputations and lives of hundreds. Some of the accused committed suicide. If the actions of Senator McCarthy had created karma, then he would have to work out that karma with every individual he ruined. This would mean that every one of them would be obligated to incarnate again just so he could resolve his karma. This would neither be practical nor fair.

The Akashic Records

The Akashic Records, "Hall of Records" or "The Book of Life" is the evolving story of life. It is an imperishable record of everything that has ever happened in the cosmos since the dawn of creation. Akasha is a Sanskrit word for "primordial," meaning original source or spirit. The Holy Ghost (the third member of the Trinity) is the Akashic

Records. The consciousness of this Supreme Intelligence is so sensitive that every vibration in the universe makes an indelible impression upon it. Unfortunately, it is impossible for thought to understand such a concept. I say that because thought can only comprehend images, form and structure. Neither the Holy Ghost nor consciousness has an image, form or structure: they are indescribable, and that prevents thought from being able to comprehend them. The only way thought can comprehend the indescribable is through metaphors and analogies. That's why the Akashic Records are commonly referred to as the "Hall of Records," a huge library without any walls, ceiling or floor.

Although the Akashic Records are considered by many to be a myth, information about them can be found in sacred documents such as the Bible, the Tanakh, and the Vedas. Keep in mind that a myth is something that cannot be substantiated by history, science or physical evidence. Consequently, there are many myths about the Trinity, consciousness and sacred texts. The key to understanding these myths or anything that conflicts with your perception is to rely upon your consciousness rather than thought. Unlike thought, your consciousness is able to embrace a concept without believing or disbelieving it. Consciousness is able to do that because it views everything as a possibility or probability as opposed to formulating an opinion or reaching a conclusion.

The Akashic Records are one of three immutable, incorruptible and indestructible standards in life. The other two are the Universal Laws and the Consciousness Absolute. It is because of these three perfect standards that life exists and is constantly evolving as it strives to perfect itself and understand itself. Yes, that is right, the perfection of God is constantly striving to perfect and understand itself. That is a concept that cannot be comprehended by thought or the conditioned mind. It is the same reason why thought or intellect cannot comprehend that life has always existed and will always continue to exist throughout eternity. Just because thought is unable to comprehend this abstract concept does not mean that it does not exist. So it is with the Akashic Records, the Universal Laws and Naught.

You can't comprehend the significance of the Akashic Records until you understand its relationship with the other two immutable standards. In other words, the Akashic Records exist because the Universal Laws and Consciousness Absolute exist. In turn, these exist because the Akashic Records exist.

Here is another way of looking at these three standards. The Universal Laws *define* life's potential, the Consciousness Absolute *develops* life's potential, and the evolving expression is recorded upon the Akashic Records. You could say that the Akashic Records are God's memory. Unlike our memories, God's memory is accurate because it is immutable, incorruptible and indestructible. The Trinity maintains its purity and integrity by constantly challenging each other – a process similar to using addition to check subtraction and division to check multiplication in math.

Your memory of what you experienced today, yesterday and throughout your life is not accurate for three reasons. One, experience has modified your perception of it. Two, you have forgotten some of it. And, three, your memory is unable to challenge itself. Only that which is not an aspect of your memory can challenge your memory. The longer your memory goes unchallenged the more corrupted it becomes and the more karma you incur. In turn, the Law of Cause and Effect causes you to experience more problems, conflict, pain, and suffering. This is why it is so important to continually question and challenge your memory.

Experience has repeatedly shown you that neither your memory nor your perception is always accurate. That is why you use various tools to examine and challenge them. Thus, the accuracy of your memory and perception is dependent upon the accuracy of the tools you use and how effectively you use them. If this is so, then surely it is wise to use the most accurate tool there is to check your memory and perception?

In addition to being accurate, the Akashic Records allow you to see cause and affect relationships that you could not otherwise see. That is important because the seeing of those relationships assists you in understanding yourself and your relationship with life. In turn, that understanding reduces the number, frequency and severity of your errors. It also prevents you from repeating them.

Outside the Trinity, there is only a handful of Akashic Record readers on earth. Just because someone believes they are able to access the Akashic Records does not mean that they have direct access to them. Likewise it doesn't tell you how accurately and wisely they use that information.

There are three types of Akashic Record readers. The majority of them are really "channelers" in that they channel one or more of the intermediaries who have access to the Akashic Records. It is the intermediary who actually has access to the Akashic Records, not the channeler.

Basically, there are two types of channelers. The first is consciously aware of what is being communicated by the intermediary. The accuracy of what is being communicated is questionable because it has been interpreted by the intermediary and influenced or colored by the channeler's conditioned mind. The channeler does not have to be psychic. For example, a good therapist can assist someone in being his or her own channeler through the use of hypnosis – these types of readings are commonly called past life regressions. At best, the information obtained by these channelers is third-hand; the process is so transparent that the reader is not aware of how many intermediaries are used.

The second type of channeler goes into a trance and turns the control of their body over to a facilitator who has direct access to the Akashic Records. This process prevents the channeler's conditioned mind from influencing or coloring the message. Edgar Cayce and Elwood Babbitt were this type of channeler.

The rarest type of Akashic reader does not use an intermediary because they are able to directly access the records. Unlike channelers, these Akashic readers obtain their information by merging themselves with the source so that they are one with it. For all practical purposes, the reader ceases to exist while they are accessing the Akashic Records. They are the most accurate of Akashic readers, and it was this type of reader who provided the life experiences in Part II of this book. Access to the Akashic Records is determined by the personality's level of consciousness. It is the ultimate gatekeeper.

Akashic readers are generally thought of as psychics because humanity doesn't realize what they really are. In fact, they're operating on a higher level of consciousness than psychics.

Reincarnation

The beauty about examining reincarnation is that it challenges your unexamined ideas, images and perceptions, especially the ones about your identity. That is important because these unchallenged images prevent you from seeing and accepting your true identity. Each incarnation is merely a different body that the personality wraps itself in to conceal its true identity from itself and others. The personality's disguise is so complete that you identify yourself as being the body.

This phenomenon occurs because thought falsely identifies itself as being the human body: our dependence upon thought causes us to identify ourselves as being the body. In turn, thought and language perpetuate this illusion whenever you identify your physical and non-physical possessions. "This is my body, my car, my spouse, my kids, my idea, etc." It also explains why you feel hurt, violated or attacked whenever someone attempts to challenge or rob the body, thought or mind of its possessions.

The personality has only one life, and it is eternal: your personality has always existed and will continue to exist throughout eternity. Each incarnation is merely the clothing the personality chose to wear in order to gain a particular experience. Like the personality, you realize that your appearance affects what you experience. People respond to you differently when you are wearing sweats as opposed to a suit or evening gown. Do you not respond differently to a police uniform than you do to a lab coat? Changing your clothing doesn't alter who you are! Just as your identity doesn't change each time you change your clothing, the identity of your personality doesn't change from one incarnation to another.

Veiled Incarnations

Normally when a personality completes an incarnation it withdraws from this realm and enters into a period of review. It reviews the experiences contained within the life cycle it has just completed and harvests the understandings achieved. That wisdom then becomes part of the personality and assists it in determining future incarnations. Once the review is completed, all memory of the actual experience passes from the personality and the only record remaining is that which is contained in the Akashic Records.

However, there are incarnations that the personality is unable to review at the completion of the experience because what it experienced is too traumatic and painful for an immediate review. And so the experience becomes veiled. Veiling has nothing to do with who the person was, but rather the trauma and emotional pain experienced during that incarnation. It is life that veils what the personality experienced.

Veiling an incarnation is actually a kindness, for the personality is still attempting to recover from the experience, and an immediate review could be so devastating that the time required for recovery would delay future incarnations.

The kindness of veiling is perhaps better understood if you examine the nature of incarnations that are so traumatic that they become veiled by the trauma. Consider for a moment those who experienced the horrors of the Nazi death camps. Imagine, if you can, witnessing your family being torn from you and being unable to save them. Would you not consider that experience traumatic and emotionally devastating? Then, too, there is the experience of a sudden and unexpected death. A person who has a life-threatening disease, or has reached an old age, is aware that death may be imminent and is therefore prepared for the experience. A person sent off to war knows that death is a very likely possibility and has accepted that as a reality. However, when death is unexpected, it is traumatizing – even more so if a betrayal is involved. I am sure by now you can see what is meant by the devastation of an experience which would veil itself.

Once the memory of the experience is veiled that memory remains within the personality, but it lies dormant and the personality is allowed to forget the experience. The personality becomes disassociated from the experience and is able to continue its journey, incarnating and gaining additional experience in the process of living. The additional experience gained by the personality assists it in gradually reaching a point where it is ready to review its veiled incarnation.

Eventually there arrives a time when the personality must become aware that there is something it needs to examine. Although veiling the incarnation served to allow the personality to continue its journey, it also has the effect of limiting the journey. There comes a time when the personality can progress no further until that experience is examined and the personality becomes free to continue.

If the veil is so complete that the personality is unaware of its existence you have to wonder how the personality realizes there is something it needs to examine. The mere presence of the veil serves as a clue to its existence, just as a locked door does. The presence of a veil and the personality's curiosity stimulates the personality to investigate it.

The personality always has the ability to simulate the veiled record and can do it whether it is incarnate or not. At such a time, a personality always has the assistance of others to come to an understanding of what transpired. Based upon the contents of the incarnation, the personality will be attracted to another physical experience that will allow it to express that understanding and turn it into wisdom.

However, it is usually while the personality is physically incarnate that the veil makes its presence known by presenting a problem the personality had not been able to foresee and plan for. The problem also provides the personality with the clues it needs to resolve it. The clue may manifest as a serious illness or some impairment that cannot be corrected. The personality cannot understand the reason for this manifestation and begins to investigate it. Each clue encourages it to investigate further. And by the time the personality completes its investigation it has experienced and reviewed what it previously couldn't examine. The process is similar to consuming a meal one bite at a time as opposed to gulping it all down at once and getting indigestion.

It may help you to view veiling as a form of procrastination. Have you ever met anyone who hasn't procrastinated or put off doing something? Each year at income tax time, thousands put off doing their taxes until the last minute. The thought of doing them stimulates unpleasant memories and reminds us of what we don't want to experience again. Just as tasks vary in difficulty, so does your ability in dealing with certain emotions.

It is only natural to avoid anything that is perceived as being unpleasant. It's the undesirable emotions attached to them that cause people to procrastinate. No two people see everything the same way or have the same emotions attached to what they see. This is why some people avoid an argument at all costs, others are willing to argue if they have to, and some like to incite arguments.

There are also those who have repressed their memories and disassociated from the experience. You can see evidence of this in people who have found their life experiences so traumatic that they have developed multiple personalities in order to continue functioning. Eventually, these and other problems have to be dealt with.

You will read about examples of veiled incarnations in the stories of Angels Mara and Anne in the Angel Stories in Part II.

Why Thought Can't Understand Reincarnation

Concrete concepts such as mathematics, logistics and physics are based upon the "known" – meaning an image, structure or some form of reasoning that can be proven. Abstract concepts such as reincarnation, karma, and the personality are based upon the unknown – meaning the absence of any image, structure or reasoning that can be proven. Therefore, it is impossible for thought – which relies upon proof – to understand something that cannot be proven.

Reincarnation is an abstract concept about the ethereal. Yet almost everything written about reincarnation is approached, analyzed and theorized as if it were a material or concrete concept. This happens because thought does not realize that the abstract is beyond its reality and that it is impossible for thought to understand anything beyond its reality. Consequently, thought is unaware that meditation – the absence of thought and knowledge – enables you to understand reincarnation.

Case in point: What is commonly called "death" is really the absence of that which created it, which is thought. That is why death ceases to exist in the absence of thought. Your ability to die to the known allows you to understand that what is called "death" is merely the appearance of death. In turn, that understanding destroys your fear of death. Such an understanding has to be experienced; it cannot be achieved by thought, knowledge or the known. You gain this experience by allowing yourself to experience the absence of thought, knowledge or the known.

Thought rejects such an idea because the knowledge it has about reincarnation allows it to rationalize, intellectualize and theorize it. This causes thought to deduce that it understands reincarnation. Thought's ability to fill in the blanks, connect the dots and create images where none exist artificially confirms its flawed deduction. Consequently, thought is unaware that it is fooling itself. For example, the word "personality" is actually an image that thought created to represent that which has no image. All the knowledge thought has about the word "personality" has no real value because it is based upon the imagery as opposed to the absence of imagery. All knowledge about the abstract is like counterfeit money in that it creates the illusion of having something of value when you don't.

Consciousness is ethereal and resides in a different field from thought, reasoning or logic. These two different fields of existence oppose each other. This is why consciousness exists in the absence of thought and thought exists in the absence of intelligence. Being whole and complete, consciousness is able to challenge everything, even itself, and so consciousness is able to observe and understand thought. On the other hand, thought – being fragmented and incomplete – is unable to observe, experience, or understand consciousness.

The attempts of scientists and philosophers to define consciousness have allowed them to discover that there are many different kinds of it. This has caused some to define consciousness in terms of awareness, others in terms of function, and still others in terms of self-knowledge. Either all these definitions are correct or none of them are. The former makes it impossible to comprehend consciousness and the latter illustrates that it cannot be defined. Thus it is impossible for thought to understand it.

Consciousness allows us to see that the basis or foundation for all thought, logic or reasoning is an image. We are able to process (read and comprehend) the words on this page because they are images. Knowledge is merely a collection of images. The simple process of adding one plus two involves images even though they are called numbers. Even the computer processes images (0 and 1). Neither a computer nor thought can process anything that does not have an image. If this is true, then how can thought comprehend that which has no image?

That is why you have to allow yourself to "let go" of what you know about reincarnation or anything you want to understand so that we can approach it freshly. You do this by meditating and acknowledging that you do not know what you think you know. When you do that, you experience an alternative to thought that is greatly superior to thought. In turn, you realize that thought is merely a mirrored image of consciousness.

Part II

The Life Experiences of Angels

Life has no secrets for those who know how to observe, question, and examine everything they experience.

The Life Experiences of Angels

The first part of this book was a philosophical and spiritual examination of the evolution of the personality. This part examines twenty actual life experiences of seven angels for your consideration. All of these angels are currently living among us today in different parts of the world – great care has been taken to protect their identities. The life experiences of each of these angels make for interesting reading and collectively they illustrate how a personality chooses various life experiences to help it achieve the understanding it desires.

In many ways each incarnation is like playing a different character in a play or movie and, as with any play or movie, there are a finite number of roles one can play. Sometimes you get to play the lead role, other times you are a supporting actor and sometimes you're a character actor. Likewise, sometimes you play the hero and other times the villain. The greater the adversary, the more heroic the hero is. Settings and environments change just as they do with plays and movies. Perhaps William Shakespeare was wiser than he knew when he said, "Macbeth, Life is but a stage."

I hope you will be as pleased as I am with the quality of the life experiences presented in this part of the book and the summary that follows each one. The Akashic Records are the source of these life experiences: they are an accurate, unbiased, and unembellished account of what transpired. The Consciousness Absolute was the source of the supporting material. I was fortunate to be able to collaborate with Luella Stroh, the best Akashic reader there is. Anyone who has had a past life reading from someone other than her will confirm that fact. That fact is also substantiated by the inability of others to access and read a veiled incarnation.

It's important to realize that this part of the book is the product of a collaboration between two entities (Luella Stroh and David Zimmer). Words cannot describe the process each of us used to access the information contained within this part of the book because words, which are physical, cannot describe the non-physical, that which has no form or image. Furthermore, it's impossible for thought to comprehend what it is like to lose all sense of "self" and become one with the consciousness that has no identity in order to obtain this information. In other words, the merger with that consciousness is so complete that you and everything else cease to exist. Even time and space would cease to exist! That made it possible to review an entire life experience of a personality in a matter of hours. Such an examination causes us to experience all the emotions that these angels experienced, just as if we had lived those lives ourselves. The experience would leave us physically exhausted and emotionally drained. Veiled incarnations were very difficult for we would have to work through the trauma these personalities experienced and because of this, it would often take a week or more to examine one angel.

Chapter Three: Angel Lee

Henry of Jamestown

Some personalities are more adventurous than others; it is as if they cannot exist without that element of excitement. That trait can be very dangerous when it is combined with naivety. And it becomes a disaster when you have been taught to grin and bear it rather than adjust and adapt.

The following life experience of Angel Lee provides a rare picture of what occurred at the British settlement of Jamestown, Virginia in 1607. It is easy to see why the sixty who survived abandoned the settlement in 1609.

This journey takes us back to early seventeenth-century England, where we find a family of prominence experiencing financial difficulties. The fortune that had ensured their position for several generations had diminished under the careless administration of the father. His unwise investments, along with his inability to resist a game of chance, could no longer be ignored. The Wilkins' family estate as well as all their properties could soon be held forfeit unless the mounting debts were resolved.

As the oldest son, it was Henry who had been summoned by the family's solicitors and made aware of the precarious position of his inheritance. He had been told that his father had ignored their counsel, and when he would no longer meet with them, it was deemed appropriate to apprise his heir of their monetary situation.

As he journeyed back to the family estate he pondered the resolutions that had been presented to him. He questioned not the genuine concern of the solicitors; their wise counsel had increased the family holdings for his grandfather. Thus, he trusted the advice that had been offered. His problem was to convince his father and brother of the necessity to determine which resolution would best serve them.

He loved his father and could not fault him for his weaknesses. That his younger brother, Thomas, was too much like his father was apparent to this young man, who feared that Thomas was ill equipped to make his own way in the world. Without the security of the family's good name, he surely would falter and fall into rapid decline. He felt sad that the strengths displayed by his ancestors had been inherited by neither his father nor his brother.

It was good fortune that found both of them home when he arrived. Deciding to delay the inevitable, he dined with them that evening – something that had become a rarity in this household. For a few precious moments, he set aside his concerns and enjoyed his father's humor as well as his brother's recount of the hunt he had just that day returned from.

Withdrawing to the library, they settled into their respective chairs to enjoy a glass of brandy. Unwilling to destroy the companionship of the evening, Henry decided not to mention his visit to the solicitors or his knowledge of the family's financial problems. Instead, a more favorable presentation had formed in his thoughts. Aware of his father and brother's inability to exercise good judgment, he decided it must fall to him to determine the most favorable resolution from those presented to him. Aware also that it would be for him to carry it through to its completion, he determined himself which of the choices open to him held the greatest potential.

As casually as possible, he began to relate a rumor he had been privy to while in London that day. Believing in the existence of precious metals in America, an association of nobles and wealthy investors was forming to lay claim to that wealth. Their plan was to finance the venture; those making the journey would share equally with the investors all profits realized. If this rumor were true, it held the promise of great wealth for all involved, he told them. To hold claim to all the gold and silver found in this new land was a most lucrative opportunity, he continued, thus the venture was being held private.

His father angered that he had received no invitation to invest. The possibility of vast riches accumulated appealed to him. He rose from his chair and paced around the room, damning his peers for ignoring him. How they could benefit from this venture monopolized the ensuing discussion, as Henry had suspected it would. When no solution could be found, Henry suggested that if they were not included in the investors, perhaps it might be possible to be included in those making the journey. Both his father and brother rebelled at the idea, saying it was rubbish and utter nonsense. Believing it unwise to appear over-anxious, he allowed the evening to come to an end with the issue unresolved.

Morning found his father consumed by the rumor. From his appearance, it was obvious that he had not slept, but had instead spent the night brooding over his brandy. Most cautiously, Henry presented his solution. He offered to return to London and request their solicitors to make inquiries for him. If they were successful, he would volunteer to join those traveling to America.

Thomas had appeared at the breakfast table just in time to hear his offer. For several minutes, all sat contemplating what he had said. Although the possibility presented was appealing, his father remained reluctant and for a moment Henry allowed himself to think that his father loved him too dearly to agree. Relieved that it was not he who would make the journey, Thomas eagerly expressed his support for the plan. It was he who finally convinced their father to allow Henry to proceed.

Henry was absent for three days. Upon his return, his father and brother hastened to hear what he had learned. When he told them of his success, it was evident that both his father and brother were already envisioning the pleasures those riches would provide. When he had allowed their imaginations enough time, he proceeded with his plan.

The solicitors had alerted him that their funds had greatly diminished, he told them. Unless they curtailed unnecessary expenditures during his absence, they would be forced to sell off some of their holdings, he informed them. If instead they acted judiciously until his return, his share of the profits would certainly secure and enhance the family fortune. The idea of altering their lifestyle appealed to neither his father nor his brother and they argued throughout dinner. With finality, Henry announced that he would make the journey only with their agreement to adhere to the advice of the solicitors. Reluctantly, Thomas and their father agreed.

Over four months they had been at sea, and everyone was sick to death of a horizon that held only more water, as well as being sick to death of one another. Three ships had embarked with a sense of high adventure. That had been quickly lost as time and weather began to affect each. Henry's thoughts at times tormented him. He desired most to believe his father had agreed to this venture only because he loved him too much to deny him this opportunity. He allowed himself to envision future generations awed by his actions while enjoying the inheritance secured by him. Quickly those flights of fancy would vanish and he would find himself questioning his own sanity. Perhaps this was a more dangerous game of chance than his father and brother had ever participated in, he thought. Was he then guilty of the same shortcomings he witnessed in them? He would hasten to reassure himself that certainly the investors were not fools quickly willing to part with their money. That they were in possession of knowledge not made known to the voyagers could only be what convinced them to finance the venture, he would reason.

When land was finally sighted they all stood and gazed as though afraid it would vanish if they were to look away, and the ship's captain led them in a prayer of thanks. Henry was surprised to see even those of questionable beliefs praying as fervently as the Christians. That all were relieved their journey was at an end was most obvious.

They continued sailing up a river searching for a favorable place to put in. It was a heartfelt cheer that went up when it was announced the search was over. The site was selected mainly because of its favorable strategic location—it would be easy to defend, if necessary. To ship their gold and silver back to England would simply be a matter of loading the ships for their return voyage, and thus it was decided to establish the settlement here rather than moving inland.

Unwilling to waste a precious moment of their time, the men quickly erected crude shelters, not much more than lean-tos. These, they were certain, would provide adequate shelter for the time they felt they would need them. Dry firewood was in short supply as everything still held the moisture of springtime. Even their eagerness to get under way did not blind them to the beauty of this strange land. A walk in the forest would reveal trees and bushes in full blossom such as never seen before; meadows appeared carpeted with flowers and one could hear a chorus of birds nearby.

What they may encounter during their explorations was uncertain; therefore to venture forth alone was foolhardy. It was decided that the men must form into groups, so from the list of names entered into the logs, the passengers of each ship were divided into two groups. Henry set off with nine other men in high spirits, certain that gold and silver were merely awaiting their discovery. They would begin their search with a thorough exploration of the area immediately surrounding them.

Their first day wore on their bodies, reminding them of the soft life they had left behind. The forests proved most difficult for walking; they were dense, with fallen timber too often in their path while unseen vines lurked in the underbrush, awaiting the opportunity to cause these intruders to stumble. They welcomed the meadow that now could be seen through the trees; surely it would be easier to search than the forest. Calling to one another, it was decided to camp here for the night and return to the settlement tomorrow while searching another area along the way.

They first sat and rested for several minutes. Separating into pairs they made their search of the meadow, disappointingly finding it as lacking in riches as the forest they had battled that day. Wearily, they opened their packs to dine on a dinner of hard tack and dried meat. Henry spread his bedroll; his whole body ached. Never could he remember experiencing such weariness, and he was eager for sleep.

The winds were calm that evening and as the moon began to rise a low buzzing sound could be heard. Within moments they were under attack: the mosquitoes arrived in swarms. The camp became bedlam as each man attempted to beat off the predators. Only by covering his head and arms with his blanket did Henry find any relief. Even then he could still hear them. When they discovered his exposed ankles, he was forced to curl into a ball that his blanket may shelter him completely. Exhausted, he slept.

The rising sun awakened them. Henry's body now cried out to be scratched. He discovered there was no end to it, for his scratching provided no relief, only the need to continue scratching. Everyone's faces and hands were covered with great welts; one had been bitten so severely his eye was now swollen shut. They had no desire to linger in the meadow and hastened to set off on their return trek.

As each of the six groups returned to the settlement, it was learned that the searches had all been fruitless. All recounted the horrors of the previous night: each had experienced much the same. Even now as they sat in their camp they became aware that they had not escaped from the mosquitoes. As he sat listening to the others, Henry found it frequently necessary to slap at himself whenever he felt a bite. Greatly disheartened, using his blanket as a shield, he lay down in his lean-to and attempted sleep.

He wondered what manner of land was this that offered no respite from these blood thirsted insects? If all his tomorrows were a repeat of last evening, how would be bear it, how could he arise each day with knowledge of what awaited him that evening? And yet, what choice was open to him? There was nowhere to go. He was a prisoner here, he realized, and he was overcome with sorrow.

With the passage of each day, Henry found his body growing stronger; it no longer evidenced his life of ease in England. His hands were calloused and stained, grime embedded beneath his nails. He could move with ease through the forests, walking for hours before resting. An acceptance of his situation had settled upon him and he was resolute in his determination to bring this to a successful conclusion as rapidly as possible.

The land remained an adversary. Summer brought heat that only increased by the day until it was oppressive. Before noontime his body would be drenched with perspiration that it seemed the mosquitoes feasted on. Never had he known such thirst, but he detested the water found here. It left a taste upon his tongue and it had an unpleasant odor to it. And still no gold to be found.

The settlement suffered losses. A member of Henry's group had started acting strangely, talking to himself and shouting at some unseen foe. His behavior continued to worsen: no longer could he be reasoned with. He finally ran, screaming, into the forest, his arms waving wildly, and was never seen again. He was not the only one to disappear: two men had left to search on their own and never returned. The others had carried one man, badly injured in a fall, back to the settlement in his group. There in his lean-to he had died alone. The soft spongy ground did not make a proper grave; water seeped into the hole that had been dug. However they had no choice but to lower him into that damp grave and proceed with the rites of burial.

His death deeply affected Henry for it made him aware that his fate could be the same. One had no way of knowing when it was their final day upon this earth, he thought. As he lay awaiting sleep that night, thoughts of home returned to him. How he longed to be there. He was surprised, however, to discover it was not the life of ease he most longed for, rather it was the simple things he missed. What he would give for water drawn cool and sweet from the well, a chair to sit in, and a bed to rest upon. His mouth salivated at the thought of fresh meat, cooked and dripping with juices. He slept fitfully that night.

Each day was a repetition of those already passed. The rising sun brought renewed hope that this would be the day that gold or silver would be found. Several times they believed it had as they dug and clawed at the rocks to reveal what lay beneath. Oftentimes it was several days before they accepted that this site offered no more riches than all those that had preceded.

The discovery of a cave caused great anticipation. Digging would not be necessary, however torches would be, to provide enough light to see. Several available grasses were tried before they had torches that burned slowly enough to allow an investigation. It was for naught – all the cave offered was cool shelter from the heat and mosquitoes. Henry was tempted to linger and enjoy this welcome relief a bit longer but could not allow himself to do so. It would have been time wasted – time better used fulfilling his purpose for being here.

Nighttimes his thoughts often turned to his father and brother, and he prayed for their wellbeing. They would soon be expecting a transfer of funds from the association that funded the expedition to them through their solicitors. He had arranged for this before his departure and had told them they could expect moneys shortly after the first shipload of gold and silver arrived in England. They had not known of the length of his ocean voyage nor the delay in locating a settlement site so he was certain they anticipated funds to be made available shortly. He had also anticipated a departure of a shipload of gold and silver long before this. Had he realized how difficult they were to find, he may have considered an alternate resolution. However many his regrets, they altered not the fact that he was here and here he must remain until he had amassed enough moneys to fulfill his promise of securing his family. It was a responsibility he had accepted and he must see it through to its end.

There was one simple act he could perform for himself, however, that may make his life here more endurable. Although it would require the sacrifice of a day, he promised himself this time. Enlisting another, they went in search of wild game. His hunger for fresh meat had been constant and he succeeded in finding one other person whose appetite was no longer satisfied with their dreary diet. Game was abundant and they decided to make deer their quarry. They could almost smell the venison roasting on the spit as they left the camp.

Much of the area was now familiar to them and they had often observed the movement of deer. Their hunt took them to where they remembered seeing them most often. Quietly they moved, slowly and ever watchful for any movement beyond them. It was not long before they were startled by an entire herd of deer, fleeing from whatever it was that had frightened them. The two men found themselves laughing at themselves and it felt so good to experience humor for a moment. They relaxed and allowed their pursuit to become a leisurely search. The afternoon sun was no longer as hot nor was the air heavy with moisture and they took pleasure in this diversion; each moving at his own pace.

Suddenly a movement captured Henry's eye. Determining another deer, he turned and motioned to his companion behind him before concealing his presence behind a tree. The deer paused a moment before something alerted it. As it startled, Henry stepped from behind his tree to take aim.

How odd, he thought, everything is moving so slowly. The deer appeared to lope gracefully off into the woods. His firearm was drifting toward the forest floor, as was he. Henry had no forewarning of his death. He did not know that his companion had already taken aim and he stepped directly into his line of fire.

Betsy Ross

We never really know our real strengths or weaknesses until they are tested. The following life experience is of a kind, gentle, determined, and strong-willed angel in Philadelphia during the Colonial days.

Historically there generally is much speculation and controversy surrounding the actions and achievements of angels. It happened to this angel. Thanks to the Akashic Records we are able to correct some of the errors historians have made concerning this kind personality.

She had taken over the household chores seven years earlier when her mother died. Although barely thirteen years of age at the time, there was no one else and the work needed doing. So, she cleaned and washed and cooked and looked after her father. He had always been a quiet man, but when her mother died he became even more so. A carpenter by trade, his days were busied with building a home or shop

for someone. He seemed happiest though when he was working on something in his own shop. There he could take his time and practice his craft as he felt it was intended. He had filled their home with the pieces of furniture that were his creations. The girl had always loved those pieces and when she was dusting them she enjoyed running her palms across their surfaces to feel the smoothness of the wood.

Her father was known and respected by many not only for his craftsmanship but also for his honesty. His good name brought him many clients, so it was never a surprise to have someone appear on the doorstep seeking to speak with him. On this day her father was working elsewhere so she informed the caller where her father could be found. He thanked her for her kindness and left.

When her father was again at work in his shop, she observed the gentleman's return. Busy with her chores, it surprised her when her father entered the kitchen to tell her there would be another joining him for his noontime meal. This was most unlike her father—it had always seemed he most enjoyed his food when left to himself.

The two men sat to the table as she dished out the food she had prepared. Her father bowed his head and silently offered his prayer of thanks to God before encouraging the other man to eat. She was not surprised that her father had little to say during the meal. His guest made several attempts to engage him in conversation, but eventually gave up and completed his dinner in silence. However when they had finished, her father showed him the furniture in the house that had come from his shop. She thought she could detect some pride in her father when the man commented on the skill of the construction.

When they went back to her father's shop, she dished out her own food. As she ate, she contemplated her father's unusual behavior: rarely would he invite another into his home, but then to display his possessions was unheard of! That it was done with pride was cause for concern. One must be ever vigilant for any indication of pride, she remembered from her schooling. Pride can threaten one's relationship with God. She knew she dare not speak of this to her father for it was not her place to question him. But she decided she could speak with God and ask Him to talk with her father.

The gentleman became a frequent noontime guest. She came to learn his name was John and that he was a furniture upholsterer. He had asked her father to construct a chair that he would then complete, and together they were working to create a special framework. Several attempts had been made before both men were convinced the completed frame was the one needed. With that, John carted the frame off to his shop where he would finish the chair.

Several weeks passed before John again knocked upon the door. Her father was again working away from his shop. Extending her father's hospitality, she invited him in for a cup of tea and he accepted. "But, you must join me," he had said, "You would not have me drink my tea alone." She thought perhaps he was teasing her, but she poured her tea and joined him. He told her the chair was completed and he wanted her and her father to come and see it. He talked on and on about the chair until she felt she had no need to see it for she already knew what it looked like. She told him she would tell her father of his invitation, suspecting her father would be most curious to see it.

Together they called at John's shop. When they saw the completed chair they were astonished by its beautiful simplicity. When John was assured that her father approved of his work, he requested him to build another frame, identical to this one. She could tell from the tone of his voice that he was excited about his idea of creating a matching set of chairs. He also wanted her father to create a small side table that would complement them.

John often called at the house; if her father was away they would share a cup of tea together before he left. She found she enjoyed his company and would blush whenever she thought of him. Chiding herself for such behavior, she would remind herself that she was no longer young, she soon would be twenty-one years of age, well beyond the age when most marry. She would tell herself that he was merely being courteous because he was a gentleman. Besides, it was her duty to remain with her father and see after him.

Again she and her father went to see the completed work. It did create a lovely setting, they agreed. The two matching chairs with the table set between them would enhance any parlor, she had told John. He then asked her father to build additional frames and tables, identical to this set, explaining that his plan was to finish them and offer them for sale as settings. Her father would not hear of it: no one would buy them, he insisted. Why would anyone purchase a setting that everyone else already owns? Who would want their home to be a copy of their neighbor's? No, he would have no part of it for it was a fool's idea, he had added, as he motioned for her to leave.

The next day she observed John enter her father's shop. Curious, she lingered in the kitchen, glancing from time to time through the window. After what seemed to be a long time, she saw John leave in haste. When her father came in to eat, he said not one word, and when he was done, he rose and went back out to his shop without even thanking her for the meal. Although she wanted to know what had happened between them, she knew it would do no good to ask, for if her father intended for her to know he would have already told her.

It was several days before her father left to work elsewhere. She was surprised when John again knocked upon their door. She was unsure if it was any longer proper to ask him in for tea, however he asked her if he may come in. Unable to respond, she stepped aside and motioned for him to enter. While she was pouring their tea, he told her he had attempted to speak with her father. She thought he meant he had tried to change her father's opinion of his idea, but she was shocked when learning she was in error. He had asked her father for permission to call upon her as a serious suitor. Her father refused, he said.

Her heart was beating so rapidly she thought it would burst. "A serious suitor," he had said! She had never had a suitor, now she suddenly had a serious one! She became aware that he was waiting for her to respond and she brought her thoughts back to what he was saying. He had asked her if she was of a mind to entertain a proposal of marriage from him. She began to weep, asking him how would this be possible when her father did not approve. John came to her and knelt before her. Taking her hand in his, he declared his deepest affection for her, as well as his admiration for her goodly regards for all people. When John suggested she speak to her father, she shook her head for she knew her father too well. She knew he would refuse her as well and feared he may forbid her to ever again allow John to enter his home.

They knew not what to do. Finishing their tea, they searched for a solution. When it appeared hopeless, John reassured her he would think upon this until he could see what to do. Both were now aware that they had entered into an agreement to marry; with shy tenderness they bid their farewells.

Try as they may, it was hopeless: they could find no acceptable solution. But, they would not be denied. After her father left one morning, she tied her personal articles into a bundle and put on her best dress. She and John stood before his minister and became husband and wife, then they returned to her father's house and awaited his return. When they told him of their marriage, he turned, walked from the room and went out to his shop without saying a word. The young woman wept that her father could not give them his blessings. Gently, John placed his arm around her and kissed her forehead. They left the house of her father and went to make their life together.

He was a good man. How she enjoyed his companionship for he spoke to her on all manner of things—

not only those of which women concern themselves, but also those that men spoke of with other men, such as business and affairs of government. She did not know that one person could know about so much. They even laughed together! She could not remember ever seeing her mother and father laugh together. She wondered if God approved of such behavior and decided that if He did not He would not have allowed her to see the mirth in what John said.

As happy as their life together was, it was their happiness only. When John took her to his family that they may come to know her, they were hospitable without friendliness. Although John invited them to their home, they never once visited. It was their only sadness. When she expressed interest in his work, John invited her to watch and, noting that another hand may be of assistance, she helped him hold the fabric as he worked. Before long she was working along with him and learning his trade.

Sadly, however, their happiness was not to last: John became ill and was taken to his bed. With work undone in his shop, she now worked alone there while also seeing to her husband. She began each day with her talk with God, and seeking His mercy she asked Him to help her find the strength to complete her day as He wished. She trusted God had sent John to her, and she knew it was not for her to know God's plan for her. If God wanted her to know, He would tell her. The simplicity of her faith carried her through each day.

John's death came quietly as he slept. The doctor had been unable to restore him to health and had explained that he would never recover. However, she was unprepared to be a widow so quickly. Not knowing what to do, she went to John's family and asked that they help her provide a proper burial for him. Never once did they express sorrow to her, however she was grateful when they agreed to see to his burial.

John was buried two days later. His family was larger than she had known–they all gathered at the home of John's parents later in the afternoon where she met all of them. Most were polite, although she felt some were acting in a manner most unbecoming: their behavior was what John had called "snobbish". She had never heard such a word until he had used it, and she remembered how she had laughed when he displayed the behavior to her. The memory brought the pain of her loss back to her anew. Politely, she took her leave and returned to their home alone.

It was so lonely! Without a husband and without family to comfort her, her days were too long and the nights offered no comfort either because the empty pillow reminded her that she slept alone. With need of something to busy her hands, she returned to John's shop and finished the uncompleted work. As she did, she contemplated how she was to provide for herself. She thought perhaps she could offer her services as a seamstress, for this would be acceptable work for a woman. She decided to make this known so she may see if anyone would have need of such a service.

One of her clients inquired about the setting of chairs that she had in her sewing room. When she related the story of how she had come to complete them, the client asked if she still did such work, for she was interested in purchasing a set. Never had she suspected people would accept a woman as an upholsterer, but she now added this as a service and her number of clients grew. She was surprised when, one day, an uncle of John's entered her shop to make inquiry. My, John was correct, she thought, for indeed this man was a snob, and he spoke to her as if she were a child who needed everything explained to her. It was apparent that this man doubted women to possess any ability to think, she decided. However, she answered his questions and he must have been satisfied with her replies, for when he departed he informed her he would return at a later date to employ her services.

When finally he did return, another man was with him. By this time, she was well informed about the politics of the times, for many a client was quick to discuss the latest news with her. Thus, when John's uncle introduced the gentleman to her, she was well aware of who General George Washington was. She was most impressed with his manners: he spoke to her as an equal as he described what it was they were there for. Seating them at a table she prepared tea.

General Washington inquired if she might have paper and pen so that he may sketch what he had in mind. As they drank their tea, he began to draw. She hoped he was a better General than he was an artist for his sketch was rough, but recognizable.

General Washington asked her if she were willing to sew this for them. She eagerly stated her willingness to do so, and he invited her opinion of the design and explained the reasoning for each section. He then confided that he was not completely satisfied, for it did not please his eye. Silently, they sat gazing upon his sketch. What it was that displeased him he could not say, so he sketched again. "It is the stars!" he said excitedly, "they do not look right. They are too perfect! Man is not perfect! Our country is not perfect! Only God is perfect. Our stars must represent our imperfection as well as that of our country. Now how do we resolve this?"

It was obvious that John's uncle was growing weary of the discussion. He rose, yawning, and stretched. When General Washington remained seated, he again sat, stretching his legs out before him and folding his arms across his chest. She suspected that he dozed off as she and the General worked on the design of the stars. The General drew several stars and they studied them. Three points were not appealing to the eye; seven were too many. However, they agreed that five points appeared to be what they were looking for. The General asked her if she could sew such a design. She felt it would be difficult to get the five points cut evenly and she expressed this to him. "What we are about to do is also difficult," he said, "It is only fitting that this should be represented in the flag that is going to represent our nation, is it not, Mrs. Ross?"

How could she respond to such a question? She could not, nor did this great gentleman expect a reply. He merely smiled at her as though already knowing she could not refuse to accept this task.

Note: Washington selected Betsy Ross because she was his nephew's widow. More importantly, he felt that her Quaker upbringing would not allow her to betray a trust. That was extremely important because Washington did not want the British alerted to how far their plans had progressed.

Now you know why the five-pointed star was selected! If you wish to share this information with others, I would suggest that you present it as a possibility or you may discover yourself in an uncomfortable situation.

Mary, South Dakota Pioneer

The next life experience for this adventurous angel takes place in South Dakota along the Missouri River, when it was known as the Dakota Territory. Fort Pierre, a hard four-day ride away, was their only protection against the Sioux Indians. Many homesteaders were unable to bear the loneliness and the harsh life of the prairie, and many found it easier to withdraw into the fantasies of the insane rather than face the reality of that hard life. The previous two life experiences of this angel enabled this personality to conquer the prairie instead of allowing the prairie to conquer her.

This examination opens with a family traveling from Ohio to the Dakota Territory, where land was available for homesteading. Horses draw a heavy wagon loaded with all their possessions; a cow trails behind. It is a family comprised of husband (William), wife (Mary) and their three young daughters. Mary's brother, George, has also accompanied them. In the midst of these travelers we find this personality has again been attracted to a female experience as the one named Mary.

Ohio had been unkind to them, Mary thought as she walked along. She hoped for a better life for her family once they were again settled on land of their own. She knew she was again carrying a child, and she prayed for a son this time: a healthy son who could help his father that William need not work as hard as her own father had.

It was early afternoon and her two youngest children were asleep in the wagon. Elizabeth, her oldest, trailed along beside her until she too asked to ride in the wagon for a while. Walking alone, Mary's thoughts returned to Ohio and the life they had left there. When she and William married, they made their home in the house of her father. William and her father had spent every spare minute of their time working on the bedroom they added to the house and, when it was completed, they moved into it as man and wife.

It was a good thing, too, she thought. Pa needed help, with George being so poorly all the time. George was six years younger than she was and was a fussy child from the minute he was born. Pa said it was because Ma died bringing him into this world and he was still crying for his Ma. With their Ma gone to her grave, Mary had taken over the cooking and cleaning. Pa's sister would come over to wet-nurse George, bringing along her own baby. Mary was thankful for the aunt, for she had taught Mary the things she needed to know. Thinking back now she realized how hard it must have been for the aunt, with a husband and baby of her own to see to. Their farm was just next to ours, she remembered, but still she left her own work to walk over and help me.

George never was able to be of much help to Pa– it seemed he was always feeling too poorly. He never looked good either: his skin was always so white and it seemed he never could put any meat on his body. He was good with animals though; even Pa admitted he could get those dang cows into the barn quicker than anybody. But, mostly, George had to sit on the porch and rest 'cause as soon as he overdid, he would begin that wheezing and have to go to bed. Calling the doctor every time got too costly, so Pa gave up and told him he had better take it easy.

Then Pa got so sick and they had to call the doctor. It was a bad year all the way around, she recalled as she walked. Hail had destroyed most of their crops – what they had managed to harvest did not bring much money. She looked at the sky and wondered if the Dakota Territory suffered the same bad weather that they had in Ohio.

Then when her Pa died, they had all the bills that needed paying. Sighing, she thought of the visit they had made to the banker. They could not borrow any more money; it had been a bad year for everyone, he told them, and now with the war and all, money was too tight for lending. Mary needed some supplies, so she stopped at the mercantile before they left town. The owner gave her what she needed, but said this was the last time. He could not afford to extend any more credit; people already owed him too much.

It had all seemed so hopeless. On their way back to the farm, they stopped at her aunt and uncle's place to see if they had any advice for them. Mary and her aunt made supper and they all talked for a long time about their money problems, the war and slavery. The children had all crawled into one bed and fallen asleep so they left them there for the night when they went home.

She remembered how she had felt when they got home. She was so sad she suspected if she were to start crying she would never stop, so she had gone to bed hoping that sleep would make her feel better. William and George had remained in the kitchen talking; she heard the murmur of their voices as she fell asleep.

When she arose in the morning, William was already out tending to the animals. When he was done he came in for his breakfast, bringing the fresh milk in with him. He was smiling like she had not seen for a long time and he ate a breakfast big enough to feed two. When he was done, he told her that he knew what to do. George had learned of new territory open for homesteading: they would sell her Pa's farm, pay off the debts and take the remaining money to tide them over until they had another crop to sell on their new farm. Mary did not feel good about the idea and she was telling William of her uncertainty when George came into the kitchen for breakfast. Before she could finish, George rushed to tell her what an opportunity this was for William. Free land, just waiting for people to stake a claim! Besides, he reminded them that this country was at war. It could be on our doorstep before we know it. What about your girls, he had asked, do you want their home threatened by those damn rebels?

So, here we all are, she thought, as she brought her thoughts back from Ohio. It had been hard to tell her aunt good-bye; she missed her even now. But at least we owe no man any money, we have our health (well, everyone except George, that is), we have some money left and there is free land ahead of us. I suspect I have to admit George was right, this is an opportunity for William, too good to pass up. But, I'll surely be glad to get there, 'cause I'm wearying of all this walking.

It was nothing like Ohio. She could have cried when she first saw their land. Where are the trees? she had asked. In Ohio we had trees and the pastures were green; here, the only trees in sight were the few growing along the river.

William had said they were one of the lucky ones because their land was along the river, which meant they had water.

Their sod house wasn't even completed when Mary gave birth to their son. Without her aunt there to midwife, William had to sit with her and help her when it was time. He told her that this new land was already providing for them, 'cause Ohio never gave them a son.

Never had she suspected that life could be so hard. She and William worked from sunrise to sunset, breaking the land. They would fall asleep every night exhausted from their day. The trip had been hard on George and he was feeling too poorly to mind the little ones so it was fortunate that Elizabeth was now old enough to take on that responsibility. With all seven of them in that little sod-house, that first winter seemed to never end. And when the wind blew, she gathered the children by the stove they had bought and wondered if she would ever be warm again. When she had asked the owner of the store what people found to burn in a stove, he had told her they gathered up buffalo chips. He roared with laughter when she had asked him what buffalo chips were for she had never heard of such a thing. Although she did not like gathering up the dried droppings, she was grateful for their warmth that winter.

Their first crop barely saw them through the next winter. With the vegetables she had stored away from the garden as well as the meat William provided by hunting, they ate better than many. They had kept the children busy all summer long gathering up the buffalo chips and weeding the garden. When the much-needed rains didn't come Elizabeth carried water from the river for the garden. She will make some man a good wife, Mary thought, for already she knows how to dry seeds for next year's planting.

The following few years were kinder and they were able to add another room with a sleeping loft, all built from the wood they bought. They had bought a parlor stove for the addition, which they used only at night when it was cold. The sod-house became the kitchen and the cook stove there was their daytime source of heat. The children all slept in the loft, and the other room was their parlor as well as where she and William had their bed along with George's. It was a might crowded but much roomier than when they were all cramped into just the sod-house.

Then came a winter such as they had never known. When the snow started, it kept on coming. Hearing of people getting lost in winter storms, William tied a rope from the house to the barn so they could follow the rope to find their way if needed. By springtime the snows had piled so deeply, Mary wondered if the sun would be able to melt it all. When finally it did, the river began rising until it spilled over its banks, flooding the nearby fields.

Unable to do any work around the farm, William decided to best use this time to make a trip into town to buy supplies: as soon as the weather got better he would be too busy to make the trip. He and George had risen long before dawn and it would be very late when they returned. "No sense you waiting up for us", he said, "better you go to bed and get your sleep".

Elizabeth had gone out to do morning chores so Mary could start the cook stove and begin breakfast. The other children had also gone outside – tired of staying inside after the long winter, they needed room to be free for a while. After getting the kettle of oats cooking on the back of the stove, she sliced bread and buttered it for them, knowing that if left for them to do themselves, she would not have any butter left. She had no desire to churn today—it would be better for her to use the day to air out quilts and do some cleaning, with the men out of the way.

She was lost in thought when she heard Elizabeth screaming for her. Dashing out to see what was wrong, she saw Elizabeth running to the river. She looked for the other children but could see only the two younger girls in the field pointing to the river. She was filled with fear as she ran to them: they were crying so she could make no sense of what they were telling her. She could see Elizabeth knee deep in water searching wildly; Lord have mercy! Her son was in the water: she saw his head bob up and his arms wave about wildly for something to grab onto. Frozen with fear, she stood watching as Elizabeth tried to reach him. The angry current was too strong; his head disappeared beneath the water. Elizabeth lunged for him and then she disappeared as well. As a cry tore free from her body, Mary ran to the water's edge. Frantic, she searched for some sign of them. She would have thrown herself into the water to search for them had not her two girls held her back, tearfully begging her to stay with them. All the day through they walked beside the river praying to find them. By sunset, she knew it was useless. They did evening chores because they needed doing. Supper was untouched: no one could swallow.

She was still up when the men returned. She had tried to think of how best to tell William his son and daughter had drowned while he was away. Because she had not gone to bed, William already knew something was wrong but he was not prepared for what he heard. He staggered as though she had struck him. Then, leaning his hands on the table, he hung his head and wept, and Mary knew not how to comfort him for she had no comfort to offer–she had only sorrow. George sat silently with them until much later, when he went to his bed. Not even aware that the fire had burned itself out in the stove, they sat in the kitchen till morning.

As word spread of their loss, the neighbors formed a search party, ordering William and Mary to remain at home. William refused: they were his children, and it was his duty as their Pa to find them. It was several weeks before the bodies were found; only when the water receded were they able to locate them. When the bodies were carted to the house they were wrapped in blankets. William tried to prevent Mary from taking one last look at her children, but she ran from him and pulled away the covers. She screamed in horror when she saw them. They were bloated and badly bruised, and she was sorry she had chosen this time to disobey her husband. Neighbors placed her children in the pine boxes that had been readied for them while others dug the graves close to the garden Elizabeth had called her own. Part of her heart was buried with them that day. It was a very long summer.

But the crops grew tall and heavy with grain that year. William said it was God's way of telling them He was sorry to cause them so much sorrow. Mary suspected it was because William spent all his time in the fields chopping wildly at any weed that dared to grow there. She observed him, as he would stand looking to the river before attacking another weed. "Perhaps it is the river he is chopping at," she thought.

Time is a blessing, Mary realized, for suddenly another spring was upon them. The confinement of winter had brought this family into closeness once again. The wounds were healed, but the scars remained as a reminder of the children she had borne and loved who were no longer with her. Sorrow is a part of life, she thought; it is with us every day we walk upon this earth. As she did every day now that the weather permitted, she walked to the graves and told her children everything that happened on the farm. She promised Elizabeth to tend her garden for her and to teach her sisters to do the same.

As with each spring, William and George made the trip to town for supplies. She thought about taking the girls and going with them, for she had no desire to spend this day here alone. She also longed to speak with another woman for a change. Winters grew so long and even when surrounded by family she had discovered she could become lonely for the companionship of another woman. But, our girls will soon be women, she chided herself, and there is much I must show them. With that she decided to remain home where she belonged. William was relieved to find her sound asleep when they returned.

George was already up when she walked into the kitchen the following morning. Thinking he must again be ill (for he never arose before them), she inquired about his health. He assured her he had never had felt better in his entire life. Then he proceeded to tell her that on this day he was heading west. Gold had been found, he had learned in town, and he was going to stake out a claim for himself before it was too late.

It was a ridiculous idea and she told him so. Their argument woke William and he joined them. She begged William to reason with him, reminding him that George was too frail to do such a thing; William only shook his head and told her to let him go. Only then did she see the valise George had packed and left waiting beside the door. Angrily, she grabbed it, opened the door and stepped outside to throw it away, only to see one of their horses already saddled. She looked to William to say something, but again he shook his head. "It is already agreed, Mary. When we sold your Pa's farm, part of that money belonged to George. We used all that money to build what we now have. It is only right that George take one of the horses as well as the saddle," he insisted. Seeing they were both set against her, she watched silently as George tied the valise to the saddle, mounted and rode off.

For days she spoke not one word to William. When she could bear it all no longer, she walked to the graves of her children and sat down to tell them what had happened. As she sat there weeping and pouring out her heart, William came and took her in his arms. "Mary, Mary, can't you see it is best he is gone from us? Are you so blind that you could not see what manner of man he is?" he had inquired.

Anger quickly dried her tears as she tore loose from him, demanding to know how he could speak so of her brother knowing how poorly his health was. William sighed sadly and said his health was only poor when there was work to be done. He told her how quickly his health improved on their trips to town, where George would head for the saloon to spend his day drinking. He was always feeling poorly the next day because he had drunk too much in town, William added. He told her how he watched George allow her to wait on him and how angry it made him. He had never spoken of this because he felt so badly about needing all the money for this farm. When George told him of his plan to leave and search for gold, he was more than willing to help him for at long last they could be free of him and his whining about how poorly he was feeling. "Poorly indeed! Did you see how he swung that valise onto the saddle? Did you see how quickly he mounted that horse? It must have been a mighty fast recovery for him to suddenly be so strong," he concluded, waiting for her reply. She was stunned! That her brother drank so when given the opportunity, she never even suspected. Sadly, she took her husband's hand in hers as they walked back to the house.

The following years were kinder to her. The farm prospered and her daughters married the sons of farmers they had known from their childhood. They remained within a short ride from her and they visited back and forth often. The grandchildren they provided delighted her and she missed them when they weren't around. It took many years for her to admit to William that life without George was much more pleasant. She no longer missed him and she enjoyed the added space once he was gone. It was a rare occasion that she found herself wondering if he ever found gold, for they never heard from him again.

She and William were buried there beside their two children on a lovely spring day. They had died peacefully in their sleep one stormy winter's night. Their bodies had been placed in pine boxes and stored till the ground was ready to receive them.

Summary

Henry's father and his brother Thomas refused to be accountable for their irresponsible acts. So Henry agreed to assume their responsibilities by being part of a mining expedition to James Town, Virginia. The consequence of his irresponsibility caused him to experience unbearable pain and suffering in a foreign land. Fortunately Henry examined his perception and life provided a hunting accident that ended this personality's pain and suffering.

The difference between the previous incarnation and that of Betsy Ross provided this personality with an opportunity to examine its perception and experience what can be achieved when you don't assume another's responsibilities. If you recall, Betsy accepted responsibility for her life when she disobeyed her father and got married. She accepted responsibility for her life again after her husband died, which enabled her to support herself.

Based upon what this personality experienced during its two earlier incarnations, you have to wonder why Mary, the Dakota homesteader, assumed the responsibilities of her bother George. To see why, you need to examine the feeling you get when you assume another's responsibilities. It causes you to feel superior to others and creates the illusion that you are in control. The feeling of being better or superior to another is very important to those who have a poor image of themselves.

The differences among these three incarnations provided Angel Lee an opportunity to see, examine and correct her perception. Unfortunately the personality valued the feeling it received when it assumed another's responsibilities more than examining its perception. Thus, it isn't by chance, accident or coincidence that this personality is incarnate today and has compounded its problems by using its education in psychology to control the lives of others. Once again life intervened to get this personality's attention and motivate it to examine its misconceptions by inflicting it with multiple sclerosis (MS).

It's important not to judge Angel Lee or anyone else, especially yourself. If you are dependent upon thought rather than consciousness, it is impossible not to be judgmental due to the limitations of thought. Your ability to be non judgmental or critical lies in your ability to rely upon consciousness and ignore thought. Furthermore, unless you know what the personality desires to experience, how can you determine what anyone should or shouldn't do? For example, Angel Lee's personality may have desired to experience MS and eventually heal its vehicle by breaking its dependence upon control.

Henry was a product of the times and of his environment. Ignorance and the desire to profit from the discovery of America created a wide variety of rumors and misconceptions about it throughout England. Being a proper English gentleman, Henry took his position as the eldest son very seriously. He was further handicapped by having been taught that it would be ill mannered of him to show his emotions. Life didn't provide much enjoyment for him. When you combine these factors with the famous British reserve, Henry's behavior was very predictable. True to tradition, he "cleaned up" the messes his father and brother created. Although he was able to accept the weaknesses of his father and brother, he was unable to allow himself the same accord. Only someone of great inner strength could have endured what Henry experienced.

Betsy was gentle and kind, but was unaware of her own inner strength. Imagine how much strength it took to marry against the wishes of her father! It certainly took strength for her to make a life for herself when she became widowed. She did have a choice; she could have gone to her father and begged him to forgive her and allow her to return to his house. That thought never occurred to her.

Mary had responsibility forced upon her at a very young age. She became her brother's caregiver and she never learned how to stop once he was a grown man. As hard as it was for her to accept his leaving, it was only then that she began to enjoy life. The many hardships they had endured were behind them; their little farm sustained them well and their lives grew peaceful.

Even though she had to grow up very quickly, she always felt a little unsure of herself or lacked self-confidence. She yielded to her husband's judgment at all times which, it could be argued, was normal for that time period. But that argument lacks understanding; it's an easy reply that is too often cited. The time period has nothing to do with it. Certainly there were many women who were willing to disagree with the men in their lives, women who had a mind of their own and were unafraid to make others aware of it. Instead what we have here is a woman who is trying to take care of everybody and ends up being used by the one she tried to help the most.

Mary loved her children dearly, and the drowning of her son and daughter was heart-wrenching. Unless you have experienced it for yourself, you cannot imagine how difficult it was for her to stand on the riverbank and not be able to save them. Likewise of her horror when she took one last look at their bodies: these were babies she had brought into the world, nursed and nurtured. She lost a part of herself when she lost her children.

Mary's false perception of herself prevented her from being aware of her strength; she would have responded differently if she had been aware of it. Her strength was displayed throughout this presentation: At age six she lost her mother and became mother to an infant brother. She traveled mostly on foot from Ohio to the Dakota Territory, where she helped her husband homestead the land while caring for her children. How could she have endured all that unless she had an inner strength to sustain her? Many homesteaders were unable to bear the loneliness and the harsh life of the prairie; for some it was easier to withdraw into the fantasies of the insane rather than face the reality of that hard life. The prairie did not conquer Mary – instead, she conquered it.

Life uses accidents, illness, and fatal diseases to get a personality's attention and motivate it to examine its perception. Humanity's lack of understanding of the nature, purpose, and value of accidents, illness, and disease causes countless misconceptions about them. It also prevents humanity from seeing that they are symbolic in that they indicate the nature of misconception and its associated behavior. For example, Multiple Sclerosis causes you to **lose control** of your body, indicating Angel Lee's desire to control the lives of others. This disease also causes you to **feel inferior**, prompting you to examine your desire to feel superior.

If the personality is unwilling to examine its perception, the illness becomes worse. In turn, the personality withdraws from the body and the physical body dies. In turn, the personality incarnates again and quickly picks up where it left off during its last incarnation. If the personality is still unwilling to examine its perception, life intensifies the illness in an attempt to get the personality's attention and to examine its perception. This process is repeated over and over again until the personality's vehicle examines its perception.

Before moving on take a few moments to examine what happened when you assumed another person's responsibilities. Try to recall as much as you can about the experience. How did it affect you emotionally, mentally and physically? How did you react? Did your response cause any problems? What was resolved or accomplished by you assuming those responsibilities?

Chapter Four: Angel John

Thomas Beckett

King Henry II came into power in England during the middle of the twelfth century. This very intelligent and highly ambitious king quickly collared his errant barons and strengthened the royal courts. He survived wars, rebellion, and controversy to become one of the most effective of all England's monarchs.

Henry, like his predecessors, desired to be absolute ruler of Church and State. The absence of a formidable adversary would have allowed Henry to gain control of the Catholic Church in England, weaken its courts, and give him absolute control over England.

The following story is about an angel whose incorruptible character kept Henry's power in check. It's an amazing story about devotion, courage, and honor.

It is the experience of a man born into a family of means in London that we first will examine. He never experienced want; his family's wealth and their position provided a childhood of privilege as well as one rich in experience. Yet, somehow, this youth managed to remain unspoiled by those around him; instead he developed an intense curiosity about his world.

Well educated, he was able to study not only in England but in France also. Upon completion of his studies, it was anticipated that he would assume his position in the family and enter into the social and political arenas of London, but he was unable to do so because he found the affectations of the rich and powerful too humorous to be taken seriously.

Thus we find he was attracted to an experience that he felt offered him the opportunity to put his education to some good use as well as create a meaningful life. He entered the service of Theobald, archbishop of Canterbury, and was assigned various church offices. His family was less than elated with his choice. In expectation of the day that he would regain his sanity, he was provided an annuity that he may maintain a proper appearance whilst also preserving his social contacts. This he found advantageous, for often his position would necessitate appearance at court; his annuity afforded him attire equal to his peers.

He now found himself in the midst of those who eagerly awaited the opportunity to enjoy the ear of King Henry II. He observed them as each attempted to improve their position. Recognizing their affectations he often was unable to contain himself and would roar with laughter, and it soon became obvious that there were those able to do damage to his position that took insult at his behavior. That he might maintain his dignity while preserving his position and sanity he became a master at offering what appeared as a compliment. Had they not been so vain they would have been able to hear the barbs concealed in his words.

His presence had been requested. It was a meeting of the king's advisers and as he quietly sat listening to their chatter he wondered why his presence was required. When a name was presented for appointment to a position of influence, he was startled to discover the king had asked for any thoughts he may offer on this. Familiar with the gentleman of whom they spoke, he was unimpressed with the choice and he responded with the skill he had perfected. As the king threw back his head and laughed the others present looked to one another in great bewilderment; they failed to recognize any humor in his response. When finally Henry regained his composure no explanation was forthcoming for his behavior, and none dared ask for one.

It was unknown to Thomas that he had been recommended to the king as one to consider for an appointment. His presence had been requested that Henry might observe him without mention of any motive. That he could provide his honest measurement of another without arousing rancor impressed Henry, and his appointment as Chancellor was forthcoming.

These two intelligent men, each delighted to encounter another able to carry on a stimulating conversation, became friends. Each found pleasure in engaging the other in a debate over some current belief or political stance. When interrupted by affairs of the day the debate would halt, to be taken up once more when opportunity permitted. That they enjoyed the companionship of one another was easily recognized at court. Evenings often found them well influenced by the wine they had supped as their bellowing laughter resounded throughout the great hall. It soon became known that to offend Thomas Beckett was to offend the king.

He not only thrived in this environment, he excelled. His upbringing, his education, as well as his ability to negotiate, all contributed to his success as chancellor. That he was ambitious was evidenced by his building projects. The strength and confidence he now found in himself allowed Beckett to display a personality that could charm another into agreement when necessary. That he was of good breeding was evidenced by the luxurious lifestyle he adopted: to receive an invite to a dinner party at his residence was to enter the inner circle of the king.

Upon the death of the Archbishop of Canterbury, Henry debated not over his successor. His choice was Beckett and he would accept no other, nor would he allow his friend the right to refuse.

Now the responsibility of leading the Church in England lay heavily upon him. The requirements of one in this position were of serious concern to him and he questioned his ability to fulfill his obligations. Much time was spent in prayer and contemplation. To honorably carry out his duties it was necessary for him to first discover that honor within himself. As he recalled the affectations of those at court he also recalled his distaste for their behavior. He wanted not to become counted among their numbers. He knew that he must find it within himself to remain true to himself whilst serving in this capacity. If unable to do so, he recognized that it would be necessary for him to implore Henry to release him from this office.

His internal struggle was waged unobserved by those around him; his hours devoted to prayer were accepted as an indication of his devotion to the church. Quietly the transformation of Beckett began until he discovered he had settled into the position of Archbishop with dignity and confidence. Those same energies displayed as chancellor were now directed toward elevating the influence of the church.

It was inevitable that these two men would soon discover their positions in conflict with one another. Henry, wishing to increase the power of the throne, attempted to gain jurisdiction over members of the clergy accused of crime. Beckett realized that Henry was using their former friendship in an attempt to gain control of the church. These two strong and determined men now found themselves confronted with the most challenging debate of their lifetime. No offer of compromise would be forthcoming from either side; no negotiations would settle the dispute, for both were resolute in their position. As the King's efforts continued, so also did Beckett's resistance, until the rift was complete.

The consequences of his resistance quickly became evidenced as Beckett, fearing for his safety, sought refuge among friends in France. Here, now separated from the conflict, he was at first able to enjoy a sense of freedom, but as time wore on the responsibility entrusted to him, and which he had forsworn to accept and fulfill, returned. To abandon his obligations was without honor, and this he could no longer accept. Beckett knew his return to England was required, and it was through the Pope that arrangements were finally made which ensured a safe return.

He would not be quieted however. Upon discovering Henry's violation of the rights of Canterbury during his absence he again became vocal in his opposition. He would not be deterred, nor could he honorably accept any diminishing of the authority of the church. His adamant refusal to yield left Henry with no alternative and reconciliation between the two men was arranged. It would be short lived.

The man who had been able to speak without arousing rancor in another no longer existed. He feared his absence had inflicted damage upon the Church's presence in England. Daily he chided himself for fearing for his own safety, for fleeing to France instead of remaining steadfast to his office. To restore honor to his position he now campaigned against those who had violated the rights of the Church as well as his rights as Archbishop, and his actions incurred the wrath of those who now discovered themselves experiencing public condemnation.

Beckett's transformation was complete; his safety he entrusted to God and his life he dedicated to His service. The Church must be the expression of God's presence here on earth, he determined, and he struggled to recognize how to bring this into realization. Isolating himself by denying audience to all visitors, he scrutinized the scriptures, seeking Divine direction.

Troubled by this challenge now confronting him, it was in prayer that he found his moments of peace. Here was where he was freed of his troubles for it was here that he yielded completely to the will of God. This allowed him to determine that the purpose of the Church was not to serve as a pawn for any ruler nor was it meant to increase the power of those who serve it. The Church was the expression of God's presence and therefore must remain inviolate of mortal influence.

Beckett's self-imposed isolation, as he sought guidance, aroused the suspicions of his many enemies within the Church. Simply stated, they now knew not what his activities were and they anticipated more reprisals. Fearing their reputations could not endure any additional challenges from Beckett, they decided to act first by arranging for the removal of this threat.

As the fading light of day began to announce time for evening prayers, he knelt at the altar. In this setting of tranquility his life was brought to an abrupt end. Murdered by his enemies, his lifeblood poured forth to become absorbed into the history of Canterbury Cathedral.

The history of the Church clearly reveals its propensity for altering history. History has recorded that it was to Henry that all looked when the death of Beckett became known, but it must be noted here that Henry was not among those responsible for this act.

Joan of Arc

The period that preceded Joan of Arc's brief career was one of chaos and despair in French history. Nearly all of northern France and some parts of southwestern France were under foreign control. The English ruled Paris and the Burgundians ruled Rheims. The English laid siege to Orléans, the only remaining loyal French city north of the Loire. Its strategic location along the river made it England's last obstacle to conquering what was left of France.

The following story is about an angel who inspired a nation during one of the darkest periods in its history.

The experience that our attention now must turn to is one that finds this personality attracted to a peasant family in France. Here we find a young girl, innocent and pure. Although the peasant's existence was difficult and often harsh, she remained a cheerful child. When their hut could no longer protect them from the cold of winter or when food was scarce, it was she who was the comforter. When small she would crawl upon her mother's lap and nestle her head to her breast, sitting quietly until her mother's attentions were turned to her. As she grew she learned to cheer her mother with song. Wild flowers or pretty pebbles were gifts gleefully presented to her.

Uneducated, her mind remained unencumbered, thus her greatest influence was her mother's devotion to the church. The most cherished memories of her childhood were of those kneeling beside her mother gazing at the statue of the beloved Madonna; it was here that this personality's devotion to God first stirred in her heart and she felt herself warmed, as it seemed to fill her up inside. This experience repeatedly visited her and she found herself awaiting its return.

Once again their meager food supply was depleted. Hoping to locate edible roots that had remained undetected or dried berries still clinging to barren brush, she set off to hunt for food. Her quest carried her far into the surrounding forest and still her pockets contained only a few molded nuts. The waning light reminded her it was time to return home, yet how could she? For supper could not be made from the contents of her pocket. Hungry as well as thirsty, she was aware that she had failed and her family would go to bed hungry that night. As she wondered where tomorrow's food would be found she began to weep, for it appeared without hope. Leaning there against a tree gnarled with age, she wept until she was emptied of tears.

Turning to begin her trek home, she heard someone call her name. Seeing no one, she blamed the breeze that had arisen. Again, her name was called out and again she saw no one. Suddenly chilled, she began to walk faster. There up ahead she could see a clearing filled with light, the last warming rays of the day's sun, she thought. Anticipating warmth, she stepped into the clearing, only to become aware that it was filled with what appeared to her as mist, filled with brilliant light that seemed to move. As she stood transfixed, she became aware again of the voice. It told her that she was loved, that God would provide for her and her family. The voice then directed her to seek out a nearby stand of brush. There, hidden by low branches, she would discover enough roots to provide food for several days. Then the voice was no more; the light waned, but the chill had left her body. Following the directions she had been given, the roots were located exactly where she had been told. With pockets full and arms laden with roots, her feet fairly flew as she returned home.

Even though she was pure innocence, she discovered herself hesitant to share her experience with anyone. Instead she told her family that she had prayed to God for his help just before finding their supper. Perhaps it was a sin of omission not to reveal how she had come upon their food, she thought. However, she had prayed several times as she searched that day and the voice had told her that God would provide and so He had, she reasoned. Kneeling in prayer before bed, she begged God for forgiveness had she offended Him in any way that day. Offering her life to Him in return for His love, she repeated her prayer of thanks for the blessings received that day.

Several days passed before she had need to return to the forest. Coming to the clearing where she had heard the voice she stopped and looked around. It was so quiet here, so peaceful. Kneeling in prayer, she again thanked God for showing her where to find food. She paused, hoping to hear someone speak to her again, but it remained quiet. Rising finally she returned to the brush, and gathered roots for several days' eating before slowly walking home.

The weather had warmed; no longer did the wind blow cold, her mother noted as they made their way to church a few days later. Soon the new plants would begin to push their way through the ground as the sun warmed the earth, her mother told her. Many others had arrived before them and their little church was almost full when they arrived. Listening to the priest lead them through their prayers, she felt the stirring of her heart and the familiar warmth fill her. When drawn to gaze at the Madonna, she was startled to see the statue smiling at her. The voice of the priest drew her attention back to the altar and when she again looked to the statue it was as it had always been.

As they walked home, she told her mother what she had seen and her mother stopped, took her daughter in her arms and wept. The Madonna had smiled upon her also, once many years ago, she told her. The first child she brought forth had not lived, she said, and when she found herself feeling poorly she feared she may not be strong enough to carry another. She had taken her fears to the blessed lady knowing she would understand a mother's heart. As she knelt, gazing at the statue, the Madonna smiled at her and she knew that already she carried another child who would be healthy. "That child was you," her mother told her. "Today our lady smiled upon you as once she smiled upon me," she said before they continued the walk home.

The ground was alive with growth once more. The girl's mother was unwell and she had gone in search of young shoots to brew a broth for her. "No, not those, child," she heard as she bent to pull the ones she had found. Startled, she straightened and looked to see who had spoken, but there was no one else present. Again the voice spoke, "Over there closer to the rock, those are the plants you seek." When she looked to the rock, she again saw the beautiful lights hovering just over the ground. She watched while they slowly disappeared before picking the plants shown to her.

She knew now that she had not offended God, He had provided as she had been told before. Once her mother was returned to health, the girl told her how she was shown where to find the plants. When her mother looked at her in disbelief, she rushed to tell about the voice in the clearing and how she had been directed to the roots. Every small detail her mother insisted on hearing. Over and over, many times she repeated what had occurred and her mother made her promise to tell her should it happen again.

And it did. Whenever there was need the voice and the vision of light would come to her. She learned to have complete trust in what she was told because the voice never failed her and she knew it was sent by God. With time, she began to have visions that appeared to take on form as they spoke to her. As they came more often she went to their priest seeking his guidance. Unable to believe that God would speak to this peasant girl, the priest knelt with her and together they prayed to God to save her soul.

There was no one. No one could tell her why God had chosen to speak to her. No one could tell her why the visions appeared or how. Because there was no one, there remained only her and God. She recalled the day she had first heard the voice and was directed to the roots. She remembered how she had offered her life to God in return for His love. My life is His, she reminded herself. It is all from Him. I must go wherever the voices and the visions lead me.

How well she did. This child, pure in thought and heart, placed her trust and her life in God's hands. Never wavering in her belief in Him, she went forth to fulfill her destiny.

The remainder of her life is well documented. It is unnecessary to recount her life any further here for volumes have been written about this peasant girl who would become known to the world as Saint Joan of Arc.

Nostradamus

The objective of this personality's entire experience on earth was to examine organized religion and its effect upon mankind. The life experience of Thomas Beckett afforded an examination of the church from the inside, as a member of the clergy. And the life experience of Jeanne d'Arc examined the church from the outside, as a church member.

The following life experience is the first of several that this personality created to achieve balance by purging these and other life experiences. It allows the reader to see how difficult this purging process can be.

Although his family was poor, the land they toiled on was theirs as long as they were able to pay the annual tribute as well as the taxes. His father's father had been given the small piece of land in gratitude for his loyalty. Since then, this proud family struggled to ensure the land would remain theirs to hand down from generation to generation.

Born the second son, he spent little time at his mother's knee. Chores were his earliest memories. First it was to find firewood for the hearth, and as he grew older he worked beside his father and older brother. Their days were long as they tended the crop they would harvest when it was ready. This would be sold to pay their annual debts; what remained would be theirs to provide for their needs.

Each year his father watched the coins in his purse disappear. They had one cow that was getting old and no longer filled the bucket when milked. He feared the day would come when it no longer had milk to give, and he hoped to be able to purchase another cow before then. But each year it seemed that there were not enough coins to do so. This year promised to be otherwise: the rains had come when needed and the field stood heavy with grain.

As they labored, cutting and bundling into the sheaves that would be taken to market, his father again noted the boulder that lay in the field. After they returned from selling the crop they would attempt to remove it, he told his sons. He had tried before, but it was too large and heavy to move without other strong arms to help him. Perhaps now his sons were grown enough that together they could move this obstacle from the field. The ground beneath it was precious soil – soil that could increase their harvest; therefore it could not be wasted.

It required all three of them to take the crop to market. Their precious cargo was loaded onto the cart that they then pulled. The sale had given them more coins than previous years and it was anticipated that there were enough to purchase the new cow when one was found. After purchasing the supplies needed for the household they returned home, jubilant at their good fortune.

Work began to remove the boulder. Previous attempts had already removed much of the dirt from beneath one side. They first located a tree that would serve as a lever, next a block of wood to place it on and finally smaller rocks, which would serve as wedges. Returning, they set to their task. The block and lever wedged into place, all three put their weight on the lever. This was repeated over and over until finally the boulder began to lift. Being the youngest and smallest of the three, it was he who ran to wedge the smaller rocks beneath lest the boulder roll back.

The work continued for several days, and inch by inch it gave way. It could be seen now that after all these years, the boulder would no more stand in their field. However this time as they pulled on the lever with their combined weight, it would not yield. Again and again they tried, but it held fast. Stopping, his father walked around the rock and realized that they would have to dig out the dirt in front as well as the rocks blocking their progress. As his father and brother set to work, they instructed the boy to run to the well for water. Thirsty for a drink, he hurried. Filling the bucket before quenching his own thirst, he returned.

His father was beneath the boulder. It had given way suddenly, pinning him underneath, his brother told him. Unable to move it, they frantically worked at removing the dirt and rocks beneath their father so they could pull him out. Their mother heard them yelling from the field and she and the other children came running to help. When they finally were able to drag him out, it was evident that his injuries were severe. As gently as possible, they carried him to the house.

It was two days before the boy was sent to fetch the doctor. As they made their way by carriage to his home, he told the doctor what had happened. The doctor questioned him about the injuries, and he answered as best he could. Upon arrival, the doctor asked him to tend to the horse as he saw to his father. Anxiously he waited, and he wondered if the doctor would ever reappear. He dared not leave the horse unattended long enough to enter the house, so he remained outside. After what felt like a long, long time the doctor came out. He plied the doctor with questions and patiently each was answered. His father, although badly injured, would recover in time, he was told. He would be unable to rise from his bed for many days, but he would return to make sure of his progress, he told him. Before leaving, the doctor even thanked him for attending to his horse and carriage with such care.

His father's progress was hampered by his worry over the cost of the doctor. It was a debt that would have to be paid, a debt that would mean no cow and perhaps not enough to meet their annual costs. If they were unable to pay the tribute and taxes they would lose the land, and then what would happen to his family? Where would they go if they lost their home?

When next the doctor came, he asked this youth about his father's progress before entering. The boy told him how his father was fretting over what was owed him and asked the doctor the amount of their debt. But the doctor told him that this was a concern between his father and himself, not a concern for a boy. Again he was given the care of the horse and carriage, a charge that by now he rather enjoyed.

This time however, he was called into the house. He was told that he would return with the doctor, there to remain as apprentice in exchange for the debt owed. Sadly, his mother handed him a small bundle that contained his other clothing before briefly taking him into her arms for good-bye. Stunned, he followed the doctor out and climbed into the carriage beside him. He turned for one last look at his home, before the carriage carried him away. Having no children of their own, the doctor and his wife welcomed him into their life. His days were spent accompanying the doctor; evenings, he quickly learned, were for his education. An entire world opened now to him as he was taught to read and write. Books were a wonder to him, although most were pertaining to medicine. The doctor marveled at the boy's mind, and realizing what his life would have held if not for his father's misfortune, the doctor felt obliged to ensure this youth's future with a formal education.

That he would follow his benefactor into medicine was never in doubt; his earliest readers had been medical books. Earning his degree, he first taught others as a means of repayment for what had been given to him. But, it did not hold the same satisfaction he remembered from seeing a body made well and whole again, and after a number of years he left the University. The doctor and his wife had both died, and left him with a small inheritance. This, combined with what he had managed to save, he decided, would be best used by setting up a private practice.

He settled comfortably into his home. A large room at the front was where he would see his patients, with an adjoining room for his office. The rooms remaining were unused except for his sleeping quarters as well as a cooking and eating area. The space was more than his simple needs required, however, the house was situated in an area he hoped would provide a lucrative clientele.

As he became known, his medical practice grew. His concern for others as well as his gentle nature comforted those seeking his skills. New patients were quick to tell him that one who he had healed had recommended him. It pleased him to learn that his reputation was one of trust, however it oftentimes disturbed his sleep as he was awakened and guided to a patient's bedside.

It had been several nights since his sleep had remained undisturbed, and his days grew longer as many now suffered from the same illness. Where he had once seen hope in their eyes, he now saw fear as people began to die. Sleep became a luxury he could no longer afford; bedside to bedside, house to house he was called. No longer did he know what day it was or whether it was day or night. He felt under siege by some unseen enemy that he may conquer should he only refuse to yield. Vivid scenes of death haunted his dreams when he managed an hour of sleep, and it became easier to remain awake.

When at last the plague yielded he returned to his home and collapsed into his bed. Fatigued beyond his own awareness, his body demanded the rest it had been denied. His sleep would remain undisturbed for several hours before the dreams reappeared to rouse him. Wearily he would then stumble from his bed to slake his thirst before returning, only to repeat the cycle over and over. Two days passed before hunger growled in his belly. How long had it been since he last he had eaten, he wondered, and how long before sleep was again a peaceful rest rather than a haunted memory?

He wed. Her father had been one of his many patients during the plague, and as he sat at his bedside she would bring him tea and remain there quietly with him. That she could remain peaceful when confronted by death's possibility impressed him, and he found himself drawn to return to this house even after her father had been restored to health. In her presence he was comforted and the dreams ceased.

The hour was late, his wife had retired before him and he now sat in his favorite chair by the fireplace. What remained of the fire was no more than glowing coals as he sat gazing at their beauty whilst pondering the many changes of his life. Slowly the colors of their glow began to rise and swirl, and, captivated by this dance, he leaned toward the fireplace that he may see more clearly. He watched as the colors continued their movement until his fireplace framed a scene before him. Entranced, he remained there until the scene disappeared. As he lay beside the dwindling coals he concluded that what he observed had been a figment of his imagination.

But the fireplace scene would return to him again and again, each time followed by nightmarish sleep. Struggling to understand what was occurring, he searched out books on metaphysics and astrology. His days were still busied with the ill, however, long into the night he could be found in his library reading, insistent that an answer be contained somewhere on the pages before him. Now, even here in this room that was his sanctuary in his expanding household, the scene would come to him and he wondered if he was going mad.

Desperate to restore sanity to his life, he sought assistance. The name of one recognized as a fellow physician gave the name of one who was an authority on the mind to him. Hesitant to have it become known that he was going insane, he allowed several days to pass – days that only confirmed his suspicions. With no alternative left to him, he decided on the next day he would visit the one recommended. Realizing how precious that man's time was, he wanted not to waste a moment recalling every detail of that scene. Hence, he took pen in hand and set them to paper. That night his sleep remained undisturbed.

Morning time he awakened surprised by an entire night of unbroken sleep. Rested and with his composure restored, he sought out this authority, taking with him his record of the dream. However, assistance was to be denied him: this authority read and reread his record but could not determine any meaning. He questioned the dreamer as to whether perhaps some detail was omitted; answering in the negative, he felt defeated.

However, again that night he slept the night through. Perhaps, he thought, it has finally passed and is over. The day remained uneventful and by evening his spirits were restored. After his family had gone to their beds, he entered his library and seated himself before his writing table. Pausing to recall the troubled man who had sat here just two nights before, a new scene began to form before him. Attempting to deny its presence, he turned his eyes to his inkwell. This only served to clarify the scene, to bring every detail into complete focus. Slumping in his chair, he laid his head upon his arms and wept, knowing the night that was to follow.

On the next day, the scene reappeared. Many nights he would attempt to deny sleep, often falling asleep at his table only to be reawakened. When he could no longer bear the mental strain, he prepared to revisit the doctor by setting to paper his vision. That night he slept once again the night through. Perhaps from this new vision the doctor will be able to offer a determination, he hoped. Again it would be denied him. It was merely suggested that he use a sleeping powder to quiet his nights for it was certain that his health would decline if this were to continue.

It was an angry man that returned to his home. Striding to his library, he located the record of his previous vision. Along with the one carried with him to the doctor, they were thrown into the parlor fireplace and set to fire. Only when no evidence of their existence remained did he turn away.

He was allowed several days of peace before it all began anew. It only served to confirm that his life was without hope; his destiny was insanity. Absent of thought, his hand picked up the pen. Perhaps if the vision is set to paper as before, a night of sleep will also follow, he thought. The morning confirmed this for him.

Now he had hope. The possibility existed that he had discovered his own solution, even if he were never to discover the reason for its onset. He established a routine: the latest vision was set to paper that he then burned, and this was followed by a night of restful sleep.

He was stunned as his visions began to play out in the events occurring in his country. Was he actually being shown the future? he pondered. How can any man know this? he questioned. No answers came to him. It would remain for him ever a mystery.

If, indeed, his visions predicted events yet to be, he could no longer destroy the records, he decided. However, for them to remain intact could result in discovery. He had no desire for his wife and children to learn of his visions, nor anyone else he knew. He feared the loss of their respect as well as the loss of his thriving medical practice. His challenge was no longer to bring an end to the visions but rather to record them in a way that would be without meaning to anyone save himself.

With the progression of time, it became increasingly difficult to record them for oftentimes their contents were unrecognized by him. However laborious, each was faithfully set to paper with the same discipline used for those which had preceded. This would then be placed with the others; with amazement he watched their numbers grow.

A new patient stimulated him to question if his records may hold some monetary value. Laughing, the patient had revealed that many of his nonsensical verses had been published, providing him added income. "People will pay good money for anything they find either amusing or mysterious," he added before taking his leave.

For several months the physician pondered what he had been told, as his number of records continued to increase. Still hesitant to reveal himself, he delayed any decision. Again, destiny intervened. The author of nonsensical verse had recommended the physician to his publisher, who was suffering from a stomach ailment. It was simple to determine its source and after treatment had been recommended the two men lingered in conversation. As the publisher confirmed what he had been told, our good doctor requested that he remain a few more minutes and excused himself. Retrieving the records from his library, he returned. He dared not pause for his courage may fail him if he did. Presenting them to his patient, he queried their value. After reading several, the publisher requested to know their source, adding that these could become a book of intrigue. This was the first person to whom he revealed himself completely. That day a trust was forged that would endure all tests of time.

The book, Les Propheties, became well known and is still known today. It remains a mystery: some scoff at the suggestion that its contents may hold predictions whilst others still study it for the hidden meanings contained in his words. That it has endured over the centuries is a testament to the life of Michel de Notredame, who is better known as Nostradamus.

Summary

Each time the personality incarnates it comes into this world with one primary asset and two secondary assets to assist it in achieving its objective. A primary asset is always will, love, intelligence, or a combination of the three. These three primary assets are difficult to recognize because they are so obvious. More often than not your perception of them prevents you from recognizing them. For example, Angel Lee's primary asset was "will," or inner strength. In the case of Angel John it is intelligence.

There are twenty-one secondary assets and they are derivatives or aspects of the three primary assets. Adaptability and naivety were Angel Lee's secondary assets; discernment and faith were Angel John's. Assets are also the personality's greatest detriment or hindrance. This can be easily seen if you reexamine any of the incarnations from the perspective of what assisted and hindered them.

The ability to recognize your primary and secondary assets allows you to maximize your potential. Equally important is your ability to recognize and realize how they can be a detriment when used inappropriately. Consequently, life is about using your assets to obtain balance.

The most experienced angels are those who are completing their last incarnation as an angel. Angel John is one of those angels and is currently experiencing a review incarnation. Let's see what you can learn that will assist you in understanding yourself and allow you to improve your own life.

Words cannot convey the simple beauty of Joan of Arc's heart. To say she was pure and innocent is employing words that are so overused they no longer mean anything. But she was innocent. She was uneducated, thus she knew only her world, a world untainted by any outside influence. She was pure. She was raised in a devout household; her whole life was God and family. Her desire and resolve to serve God was so absolute, nothing could prevent her from doing it. It is impossible for thought or intellect to comprehend the wisdom that was required to be attracted to this vehicle and to ensure that it achieved its objective. Consequently, Joan of Arc inspired tens of thousands of her countrymen to do more than they thought they were capable of doing. Likewise, it is impossible for thought or intellect to comprehend the profound love so many personalities have for her.

The life experience of Thomas Beckett would require that he be educated, both academically and streetwise. This incarnation would also require social status and connections that would gain the attention of the king. These requirements attracted him to be born into a family that could provide the necessary education, wealth, and position. Beckett's background and licentious appetite for wine, women and folly ensured that he and Henry would be attracted to each other.

Beckett's overpowering sense of honor and his relationship with Henry waged a tremendous internal struggle within him and caused him to seek Divine direction in his final days. His discerning abilities enabled him to determine that the Church was not to serve as a pawn for any ruler, nor was it meant to increase the power of those in service to it. The Church was an expression of God's presence and therefore must remain inviolate of mortal influence. That realization was so clear and complete that nothing could stop him from fulfilling his destiny – not even the fear of death.

Beckett's understanding of the role of the Church was profound and revolutionary. If you asked yourself "What did Beckett mean by that?" you will see that his position was in direct opposition to why the Church of England was established in the first place, as well as any other church.

The life experience of Nostradamus (Michel de Notredame) is basically two life experiences in one incarnation. Destiny abruptly interrupted the life experience of this successful doctor and author.

The purpose of the interruption was to prepare this personality for its final incarnation as an angel and its greatest challenge.

It pleases me to be able to add some clarity to the life experience of Nostradamus because it has been greatly misunderstood by most people, especially the personality that experienced it. Although many secretly wish they could be as great a clairvoyant or prophet as Nostradamus, few personalities have experienced the turmoil this man endured because of his visions. For a long period he constantly questioned his sanity and feared he was going insane. What a joyous relief it was for him when his ability to observe provided him with a method of freeing himself from the visions that haunted his sleep.

The most predominant and apparent purpose demonstrated by Nostradamus was discernment. It required great discernment to determine the contents of the visions as well as to record them as quatrains. Often today this is mislabeled as good judgment; however, discernment is totally lacking in any form of judgment. Rather it is the ability to determine truth, an ability that increases as one's trust in self increases.

What historians and those captivated by his quatrains fail to realize is that he had the wisdom not to write them so that the average man could understand them, yet make them appealing and mysterious. Many incorrectly deduced that he wrote them that way to protect his reputation as a respected physician and astrologer, and others figured he did it to preserve the life he had established for himself. Most reasoned that he did it to protect himself from the wrath of the Inquisition. Surely a seasoned personality with his experience was fearful neither of controversy nor death? Only a few realize how his visions would have impacted society and would have radically altered history if they had been easily understood.

How would people respond if, year after year, everything happened as he predicted? How do you think the public would have acted if on September 10, 2001 they "believed" that thousands would die on September 11, 2001 in New York City? What purpose would be served if you or the personality always knew what was going to happen before it happened? What would you or the personality learn?

There is one more important factor to consider that greatly impacted Nostradamus' life that has never been mentioned. He observed a past life of his that troubled him because it involved making an extremely difficult decision that greatly haunted him. Observing that life stirred powerful emotions which he buried deeply because he couldn't accept what he experienced. If that wasn't bad enough, he witnessed his present incarnation and that was equally troubling and disturbing. These two events caused a very decisive personality to be an indecisive one, especially today. Fortunately, life has intervened and is assisting him in resolving this issue.

The life experience of Nostradamus clearly illustrates how your assets are also your biggest liability or shortcoming.

Have you taken time to determine what your primary and secondary assets are? Why wouldn't you want to discover your greatest resource? Ask your friends and relatives to assist you in identifying your assets. They are probably more aware of what they are because they see you differently than you see yourself. Keep in mind that your shortcomings or faults are also assets. With that in mind, even people who dislike you can help because they have a tendency to point out your faults and shortcomings.

Chapter Five: Angel Lyn

Pastor John

Between 1609 and 1664 Dutch entrepreneurs established New Netherland, a series of trading posts, towns, and forts up and down the Hudson River that laid the groundwork for towns that still exist today. Fort Orange, the northernmost of the Dutch outposts, is known today as Albany; New York City's original name was New Amsterdam, and the New Netherland's third major settlement, Wiltwyck, is known today as Kingston.

This life experience is of an angel who helped settle New Netherland. It is a story of the human spirit, the innate desire to live in freedom no matter the difficulties encountered. It is a story of unconditional love in that this angel had little, if any, concern for self. Although his wife and children were important to him, he did not cherish them beyond all others. He treated everyone fairly and equally. It is also an example of acting upon faith in your own inner guidance and in God.

This personality was attracted to experience the physical realm as a male born to a family in the Netherlands early in the seventeenth century.

It had been a journey fraught with hardship. He had anticipated they may encounter storms, for they are frequently encountered at sea. However, the illness that fell upon them arrived unforeseen. There had been several sea burials, including this personality's stillborn infant. Only through God's tender mercy had his wife, Matilda, survived, although she remained weakened from the ordeal. Their young son, his namesake, had been unaffected and they offered prayers of thanksgiving for the boy's safety.

It was a weary shipload of souls that welcomed their first sighting of the new land; so often they had feared this moment might be denied them, and now that it was finally upon them, they gathered excitedly at the ship's railing. Softly he began to sing a hymn familiar to all; one by one they joined with him, rejoicing in the knowledge that they approached a safe harbor.

As they disembarked most were startled by the primitive conditions that greeted them. It was known that only in recent years had people begun to establish villages here, but still they were unprepared to see the crude dwellings surrounding the harbor. They were quickly confronted with the reality of their new homeland: the life they would have here would be of their own creation.

His flock gathered around him looking for guidance. They could not know that suddenly he felt as lost as they, but he knew that he must be strong and must offer them hope and encouragement. From somewhere he remembered a prayer taught him by his father, and, reciting it from memory, he prayed his voice would not reveal the misgivings in his heart.

It was weeks before the group located a site where they could begin to build cabins to shelter them. Already they had been told of winters past, of the cold winds and snows that arrive with sudden ferocity. Everyone toiled throughout the light of day, working together to complete one cabin before beginning another.

Seven cabins stood when it became necessary to cease their labors until spring. The group had insisted that John and his family reside in the largest cabin, where they would all worship until their church was built in the springtime. The remaining cabins were shared; each housed two families.

The winter was kinder than they had been warned–many days the sun shone and the men worked to fell trees in preparation for springtime. The children welcomed the freedom of the outdoors and helped by gathering fallen tree limbs for firewood. When forced to remain indoors, John busied himself with building an altar for their worship, believing it would serve to remind his flock of the future they were building. He told no one of his project, and it remained covered when he was not working on it. All through the winter, he carved and polished until it stood ready for worshipers, and when it was finally revealed to them, they wept when they saw this symbol of their faith. After dedicating the altar they marked off the site in the snow where the church would be built, placing a cross made of tree limbs where the altar would stand.

By springtime they had become an organized community. With enough trees ready to resume building, those most skilled in carpentry set to erecting more cabins, while another group of men continued to cut trees and prepare them for building. The women and children worked on garden sites, digging and planting, to ensure that they would have an ample food supply for the coming year. Slowly the little village took shape.

"Oh, truly we are blessed, for Matilda again carries a child!" he announced proudly. Their son, John, was growing and anxious to be away from his mother's knee. A daughter would be most welcome, for although women appear proud when bearing a son they appear to find pleasure and comfort in raising a daughter, he had observed. A daughter must offer them companionship, he realized, the same companionship a father enjoys with a son. Well, be it son or daughter, it is up to God for He knows best what is needed.

With sufficient cabins now standing, each family's attention turned to the construction of their church. They dedicated the site to God before sitting down to a feast of celebration; it was the most pleasurable afternoon they had enjoyed since their arrival. As he watched Matilda move about to refill empty plates it became obvious that her time must be near. She stepped slowly, pausing from time to time as she eased her back. It filled him with gladness to see one of the other women take the kettle from her hands and lead her to a place where she could sit and rest.

Their second son was born that same night. Their oldest, John, bore his name, and this second blessing they named Matthew. He arrived healthy with a strong set of lungs. Their community grew by three more that summer, all boys.

When the church was completed the altar was moved from John and Matilda's cabin. With much ceremony, the men set it in place before looking to John to lead them in prayer. As he stood among them he was overcome with emotion. The scent of the trees still lingered and he could hear the birds singing outside the doorway where the sun was beaming through. He could feel the presence of God and it was profound. When at last he was able to speak, he thanked God for the many blessings they had received and asked for His continuing guidance in all their labors.

Everyone believed the first worship service in their church was special for it marked the baptism of the four sons born to their community that summer. The gardens had flourished, providing an abundance of vegetables to accompany the game that was hunted in the nearby woods, and they enjoyed a grand feast after the ceremony. Even the children contributed: berries had been discovered growing wild and the young girls had picked enough for all to eat their fill.

"Man cannot ask for more than what we have received," John thought; the rich land offered sustenance beyond belief. Already word had spread of the craftsmanship of their cabins and two of the men had been hired at fair recompense to erect a public building in nearby Boston. The hardships endured had been met with great rewards; the homeland left behind was drifting into merely a fond memory.

Winter came again, bringing the bad weather they had been warned about. "These are the snows they spoke of when first we arrived," John thought as he surveyed the results of the latest storm. "How fortunate that we were spared this experience until we were properly sheltered with an ample wood supply nearby." The cold weather brought with it an illness that settled upon the children and, although it seemed not to be serious, it did keep the women busy attending to the sickbeds. Matilda had both John and Matthew stricken at the same time and she prepared kettles of broth for them as well as the others.

Visitors were infrequent to their village and it was not until travel became possible once again that a strange family arrived. Offering them the hospitality of their home for the night, Pastor John was troubled to discover they planned to relocate to a distant village. Why had they not waited to travel when winter no longer threatened, he had inquired? It was foolhardy to endanger his wife and child by journeying now when a storm could find them without shelter, he had added.

Yes, the man agreed, if he could have chosen to remain and travel the roads in springtime, his journey would have been safer as well as easier. However, he could no longer remain in the home he had built and worship God according to his beliefs. The Massachusetts Bay Colony had expelled him and his family for refusing to accept the beliefs of the Colony. He was taking his family westward, where he hoped to find the freedom he had sought when bringing his wife to this land.

They would not make their journey in winter, Pastor John insisted. Their cabin was large enough to shelter all of them and the family would remain until it was again safe to travel. Matilda welcomed another woman into her presence and the two quickly became friends.

When springtime came the settlers discovered that even they were no longer welcome here: they too were given the choice to either return to their homeland or be expelled from the Colony. Greatly disheartened, Pastor John called a meeting of all the men in the village. After leading them in a prayer for guidance, he haltingly told them of the decision they were being forced to make. They must abandon the homes they had built, the gardens awaiting spring planting, if they were to continue in their own beliefs.

Not one man spoke; the silence echoed their disbelief. They had traveled across an ocean, endured hardship and illness, believing they journeyed to freedom. When it became apparent none were able to respond, Pastor John began to speak again.

"We embarked on this journey together. We have melded into one family through the sharing of our experiences. Therefore, if it is your decision to return to our homeland, Matilda and I will return also. There would be no fault in that decision for it appears this land no longer welcomes us. Should you decide to move westward, Matilda and I will join with you also."

They discussed the problem for many hours, seeking any information Pastor John had learned from the couple that had wintered in his home about the lands westward.

He had little to tell them, however it was known that
the Massachusetts Bay Colony did not extend that far.
Some argued that they might move only to discover yet
another restrictive governing body. He could offer them no
reassurances; they would have to act on faith alone.

When they were finally able to accept that there were
only two choices—move on or return to the Netherlands—
they realized they must move on. Some felt it would be
wise to first send out a party to search for new home sites;
most would then remain there and begin building while
two returned to lead the others. John was troubled by this
suggestion, although he could see that there was sense to it.
They had begun this journey together; everything they had
encountered had been as one family. When he expressed his
feelings to the others, they agreed: they had arrived together;
they would leave together!

The move would be less difficult than when they first
came. They had built three wagons, and had horses and
mules to pull them. Before the loading began, the men had
assembled in secrecy. Puzzled, John watched as they walked
to the church, for they had not asked him to lead them in
prayer. They reappeared carrying the altar he had built,
and with great gentleness it was loaded and covered with
protective quilts. He was so choked with emotion he could
not speak. The altar would not remain behind; he had not
realized how much it symbolized to his flock. That it took
up precious space mattered not; they would choose to leave
something else behind rather than abandon the altar.

It was heart-wrenching to watch as they closed the
doors on their cabins for the final time. Some walked out
without looking back while others paused, touching the door
as though saying good-bye. John knew each was left as clean
and tidy as if visitors were expected any moment. The garden
plots were readied for planting; already it could be seen
that seeds which had fallen during last year's harvest were

beginning to grow. From somewhere in the forest he heard the call of a dove. It was time to move on.

The first days of their journey were solemn; most appeared lost in memory of the homes they had just left behind. It concerned him that even Matilda spoke only when necessary and was often at a loss for patience with their sons. He at first believed it would pass, however after several days it became apparent that it must be addressed.

They would stop early tonight, he told each of the wagon drivers, explaining that after they had eaten they would gather for a meeting.

"Today was another day for which we must remember to give thanks. Fair weather has been with us on this journey; our animals have encountered nothing too difficult for them to overcome. Truly God is guiding us. We cannot know His plans for us, but He has never failed us. Come with me, let us give thanks."

He led them to a small clearing they had passed before making camp for the night. None except one knew what awaited them there. As they stepped into the clearing they saw the altar standing where he and the wagon driver had placed it. John could feel the moisture in his eyes as his entire flock knelt and looked to him to lead them in prayer. "Thank you, Father, for showing me that they merely needed a reminder," he thought, "for truly you are with us every step of the way."

The morning sun rose on a new day for all of them. They had given their burden to God and their hearts were lighter. It had been decided to journey far beyond any possible reaches of the Colony, and that would mean added travel. Despite this, laughter could once again be heard.

They settled in New York, where they built the lives they had dreamed of as they journeyed across the ocean. Again, cabins were built first. When the church stood ready, a ceremony was held to set the altar into place. It stood as a polished reminder to each of them, a symbol each had come to cherish.

Teacher William

Attracted to return to New York early in the nineteenth century, this personality chose once again to experience the physical realm as a male. Born into a family of means, this personality was seen as the "black sheep" of the family – the one who everyone was certain wouldn't amount to anything. Neither this personality nor those who adversely viewed his independent nature could have imagined how life would provide opportunities for him to successfully express himself. Often what we don't want or what is seen as undesirable is really a blessing in disguise.

As in previous incarnations, this personality listened to and followed its inner guidance. Few are willing to march to a different drummer because of how they are seen and the criticism they receive. Once again this personality never realized what a kind, generous, and loving person he was. The first to realize it was his grandmother, Nana. Later his wife, Nora, and his children knew. Eventually, hundreds of others realized the worth of this brave and unselfish individual.

"Christmas time in New York City! How could anyone wish to be elsewhere?" he wondered. Home from college for the holidays, he was enjoying shopping for gifts for his family. Already laden with packages, he paused to examine the festive display in a store window and observed James, his older brother, inside making a purchase.

He decided to move on rather than allow James to spoil his festive spirit. He was positive he was purchasing his wife's Christmas gift. He was also positive that the gift had been given most serious consideration and that it would prove to be a very wise investment. His brother never did anything without first determining the merits of such an action–whether the dollars spent would bring him a substantial return.

"Spare me the life of a banker," he muttered to himself.

From early childhood, William had been constantly reminded of his brother's achievements, for James excelled in academics and was always a model student. William's own grades were always a source of disappointment to his father, as well as his deportment. Too many times his father had been summoned to the school because of his involvement in a schoolyard fracas. His grades as well as the reports of his behavior had prevented him from attending Yale, where James had graduated with honors.

Only his grandmother's gift remained to be chosen. The gifts for the others in his family were always traditional, but for his grandmother it was always something whimsical. She was his dear Nana and he knew she adored him as well. At times he felt as though they were co-conspirators as they attempted to instill some life into the stuffy family dinners. How her son could be so opposite to her was a puzzle.

He searched for something that he felt would delight her and bring that special glisten to her eyes when she opened the wrappings. He discarded several ideas as nearing the traditional; that simply would not do. Deep in thought, he had wandered into a shop that was new to him and browsed about in fascination. Remembering her joyful laughter when he had presented her with a bird two years past, he knew he had found the perfect present when his eyes fell upon a sketch of a bird. The artist's name, Audubon, was unknown to him, but it would make no difference to her–he knew she would love it.

Christmas Eve had been observed with the usual party his parents hosted each year for their friends, and on

Christmas morning the family exchanged their presents before attending worship services. As he had anticipated, his grandmother delighted in the sketch he had found for her. It humored him to observe his brother glance at the artist's name before offering his comments.

Dinner was over and William knew he was mere moments away from being summoned to the library where he would receive another of his father's lectures. He did not look forward to sitting and suffering the sharp rebukes his father could hurl when angered. He was certain his father had been notified that his grades had fallen. It would take a miracle for him to graduate. Running his hands through his hair, he wondered why he always managed to irritate his father so. He knew he could raise his grades if he wanted to, if he would attend his classes instead of roaming the countryside, but the fact was that he simply did not find textbooks of interest; they just put him to sleep. And to sit through a lecture at college! He'd had enough lectures at home to last his entire lifetime.

Never had he seen his father so agitated. He paced the library with rapid steps, pausing only to deliver another diatribe.

"You are doing this on purpose to embarrass your family." He shot at the young man.

Now isn't that ridiculous, William thought. Yes, every morning upon awakening my first thought is "What can I do today that will bring further embarrassment to my family?"

"Whatever it is you are up to, I will tell you now that I will not have you in the bank. If you believe I will bring you in there to cause your brother added dismay, you are in error. James earned his position at the bank; you have earned nothing but failure."

Thank you, thank you, he thought. Spending the rest of my life in that bank would be the same as condemning me to hell. I may be a failure, Father, but I am a lovable and happy failure. He had to conceal his smile as he pondered what his father would do if ever he spoke his thoughts aloud.

"I don't know what to do with you, William," he said with a heavy sigh as he seated himself behind his desk. "I would be throwing away my money to return you to school. You'll remain here until I can come up with a solution—perhaps your brother may have an idea or two."

Oh, knowing James, I am certain he will have an idea. Nothing to my liking, however, William realized.

When next summoned, he found James standing beside their father in the library. *How similar they are, he thought, James even stands as Father does, with one hand in his jacket pocket while the other toys with his pocket watch.* Refusing to be intimidated, William strode to the fireplace and nonchalantly rested an arm on the mantle.

After lengthy discussions about his future, they had arrived at two possible solutions to his problem. It would be his decision to make, he was told. The first solution presented was that of marriage to a cousin of James's wife. It would be a good marriage, they insisted, for she came from a notable family. William knew his father had been made guardian when her parents died, managing her accounts and investments. *What he means by a good marriage is that she has wealth to make her attractive,* he thought.

His second option was a novel one and he wondered who had thought of this one. He was surprised to discover his father had property in Ohio, and even more surprised to learn it was the result of a failed business venture. *Apparently not all your investments earn you a profitable return, Father,* he thought in amusement. If he chose this solution, he would travel to Ohio to develop the property. His father would finance the venture for five years; after that he would have to live on income earned off the property or by his own labors. Perhaps experiencing frontier life would help him appreciate the life offered here in the city.

William was given 24 hours to consider his decision. Without a doubt, his father would summon him again at the end of the allotted time, when he would be asked for his answer. Neither option appealed to him; both appeared to be banishment. He was either banished to live a life married to a woman totally lacking in humor or to live a life lacking in civilization.

Bless Nana; she could always be counted on to see what others could not. He had taken his troubles to her as he had done for as long as he could remember. She was never harsh, never chastised him. In her own gentle way she always found a way to encourage him. This time she suggested that perhaps what he needed was an opportunity to be his own person, away from the influence of his father and his brother.

"You may find Ohio to your liking. It certainly will require one with your sense of imagination to turn a failed business venture into a success," she said.

His decision astounded James – the look on his face revealed he had anticipated a wedding. Their mother cried, as William knew she would, while his father stood stoically by. He was reminded that it was his decision, and he would have to live with it. Nana winked at him while a naughty smile played upon her face. With a display of bravado, he bid them farewell, telling them he was setting off on an adventure few men are given the opportunity to experience.

Ohio was not the frontier his father had depicted, although not what one would call civilized either. His father's property was located near a place called Marietta—a town which could only grow, for he doubted it could be any smaller. As he investigated this hamlet, he realized he was actually enjoying himself. *This really is an adventure*, he thought, *who knows what awaits me here?*

Nana had been correct in her prediction, for he did like it here. He found the people friendly as well as helpful, and he enjoyed his new-found sense of freedom. No longer was there someone waiting for him to make an error, waiting to compare him to his brother. Here, he was his own man.

However, the question of what to do with the property remained a puzzle. He had written to his father, suggesting the property would become more valuable with the growth of Marietta. He had no desire to farm the land, but it made no sense to leave it unused. Perhaps they could run some range cattle—use the land for grazing, he had added.

By return mail, he received his father's agreement; he even congratulated William for recognizing the future potential of the land. Yes, by all means, go ahead and run some cattle. Build a small house on the property if necessary and perhaps a shed, he suggested. Of course, there was also the reminder to keep an eye on the costs.

Walking to the bank, he realized he had become a respected member of the community. He recognized every person he encountered and they all knew him by name, calling out their greeting to him. He did not enjoy his visits to the bank; they brought back memories he found unpleasant. However, the financing his father provided was sent here and deposited into an account for him, so he could not avoid these visits.

Are all bankers without humor, he pondered as he waited for the little man? They must be born that way, he speculated, remembering his brother.

He withdrew the funds needed to complete the shed he was building. The banker had engaged him in casual conversation, made polite inquiries about his cattle and how the shed was coming along. In an attempt to introduce some humor into the conversation, William remarked that his college education had not included carpentry. The banker had merely raised his eyebrows and asked which college he had attended. Typical, William thought, facts and figures are all they find of interest.

He had been looking to buy another horse, considering the possibility of adding horses to the grazing land. Perhaps even breed horses, he thought. He knew nothing about horses other than how to ride one, but did not believe it could be too difficult to breed them, for the horses knew what to do. "A house, shed, cattle and horses, a real country gentleman I'll be if I'm not careful," he thought with amusement.

William saw two riders approaching–not a common occurrence, although not a rarity either. When they neared he recognized both of them and called out a welcome. Dismounting, they wasted no time in stating the reason for their call. The banker had made it known that he was an educated man. Marietta was growing, and the town needed a new schoolteacher. The last one had gone back east where he could make more money, they explained, before asking him to accept the position.

"Don't pay much. Town's too small yet to offer a man a livable wage," they added, almost as an apology.

The idea appealed to William, and thus he added teacher to his growing list of accomplishments since his arrival in Ohio. None of them turned a profit, but the life he was building for himself provided a satisfaction that would never appear on a bank statement. His first days of teaching were awkward, as he found himself modeling himself after teachers he remembered. It was only when he relaxed, deciding to fashion his own style of teaching, that he discovered he enjoyed working with the children.

The five years were up, and William's father wanted an accounting for the financing he had provided. William had written to him, describing in detail the improvements that had been made to the property. There were no profits realized as yet, neither from the cattle nor the horses he was raising. He believed the town would continue to grow and he exaggerated a bit when he wrote of the new enterprises moving into the area. In closing, he mentioned his teaching position, adding that he would be interested in purchasing the property should his father decide to sell.

The message simply stated that his presence was required at the bank. Its brevity did nothing to still the fears he imagined. He suspected the worst: that his father had sold the land to some speculator to recover his investment. He was unprepared for the messages he received.

Nana had died. They had not been able to notify him in time for him to attend her funeral, and for that they were sorry, for they knew how much he loved her. She loved him equally and had remembered him most generously in her will. He must notify his father if he wished the funds liquidated and sent to his bank in Marietta or if he wished to leave the money invested where it was.

Stunned, he sat lost in memories of that gracious lady. He recalled the letter he received after sending her a buckeye tree for Christmas his first year here. The family had thought it his most outrageous gift to date, which delighted her even further, and she had insisted the tree be planted in a place of honor in front of the mansion.

William became aware the banker was continuing on, as though he still had his attentions. He straightened in his chair, running his hands through his hair in an attempt to clear his mind.

I must have misunderstood or heard incorrectly, he thought, begging the banker's pardon and asking him to repeat what he had just said.

"Your father has signed the property over to you: it merely requires your signature here," he repeated, showing William where to sign.

Overnight his fortunes were altered. He was now both a property owner and an investment holder, for he decided to leave Nana's investments where they were, at least for now. His teacher's salary was enough to provide for his simple needs.

Just as quickly he became a married man with a family. His wife was a widow – she and her husband had farmed the land next to his. When the husband died from consumption, William had paid his respects and insisted she let him know when she could use a helping hand. Her two sons attended his school, and after their father's death they were absent all too frequently. Concerned about their education, he began dropping by to tutor them. He was always made welcome, and usually ended up staying and enjoying supper with them. Even though life could not be easy, the boys' mother maintained a sense of humor that he found attractive. It felt natural and right when he finally asked Nora to marry him. She and the boys moved into his house, and William believed his world was complete.

Their family grew. After the birth of their second daughter, they added two more rooms to the house. Nora wanted a front porch, so he added one! He believed there was nothing he wouldn't do for her and the children. William proudly took them all back east to visit his family. It was another of his adventures, traveling with a wife and four children. He was pleased that they enjoyed themselves, and his parents were pleased by the visit. His father refused to allow them to leave until William assured him this would not be their only visit.

William was late returning home one evening. A lamp still burned in the house but he knew all were in bed, sleeping. He had business in town, and had eaten supper there before meeting with the school board members. He was encouraging them to look for a new teacher because his own interests took up his time more and more. When he attempted to convince them of this, they believed he was merely trying to negotiate a higher salary and when the hour became late, they brought the meeting to an end without resolution.

He unhitched the horse and led him into the barn. Lighting the lantern, he picked up the bucket to fetch feed for his animal. Hay suddenly drifted down from the small loft and he wondered what manner of critter had established residence there. He froze when he heard a faint moan—it was no critter up there.

What the hell do I do now, he asked himself. Quietly setting down the bucket, he picked up the pitchfork, crossed to the ladder and climbed up. What his little loft held was beyond anything even he could have imagined.

Runaway slaves! One look was all he needed and he knew. Sure as hell, they were runaways. Two of them and a baby! Lord, oh Lord, why me, just when things were running so smooth?

They were cowering in the hay, their eyes shining with fear, "Please, please, mistuh, don't send us back," they begged.

The pitchfork had fallen from his hand. Now he slumped to the floor of the loft, running his hands through his hair. What to do, what to do? They couldn't stay here, he couldn't endanger Nora and the children. On the other hand, no way could he send them back, but what was he going to do with them?

"I won't send you back, I promise. Just let me think what we can do with you, where you will be safe. Are you hungry? Of course you are. Stay here, I'll fetch you some food."

He hurried to the house, knowing Nora would have a plate of supper waiting for him. He grabbed the plate along with a loaf of bread and turned to go back to the barn.

"William, whatever are you doing? Where are you going with that food?" Nora called out.

His entering the house had awakened her; now she stood before him awaiting his answer. He was never any good at lying–it always came out sounding like the lie that it was. In a rush of words he told her about the runaways in the loft.

"Nora, we can't send them back, we simply cannot do that," William pleaded.

She agreed. And when he explained he did not know where they could be safely hidden, she knew exactly where to take them. To her old house! William was holding the property for her until her sons were old enough to farm the land. Now she told him of the cellar concealed beneath the shed. It had once been used as a house, she quickly explained, before a regular house was built. The cellar was long forgotten; hadn't been used for years.

They carried the food to the barn; Nora remembered to bring milk as well. After the runaways had eaten, William led them across the countryside in the moonlight. The door to the cellar was difficult to find and harder yet to pull open. It was musty smelling and damp in the cellar, but they would be safe there until he found a way to move them. Leaving candles for them to burn, he warned them to remain there until he returned.

The next day William and Nora considered every possible idea they could come up with. They would have to be moved further north–he had heard tell that runaway slaves were safe there. By nightfall, their plan was ready. Again he crossed his land, carrying a sack full of food along with the syrup pail of milk Nora had filled. As they ate, he told them the plan, urging them to hurry.

It was bold, but he believed that's why it would work. When they returned to his place, Nora had the wagon ready to go. Quickly, they climbed on and lay down while William covered them with hay. Tying his saddle horse behind, they set off. He drove until the sky began to lighten, and then pulled over into some trees. Nora roused their cargo while he retrieved the rope from under the seat.

"I'm sorry, but this has to look real," he apologized to the couple as he wrapped the rope around their waists and secured it to his saddle. He tied the man's hands, but left hers free to carry the child.

"God be with you," Nora told the runaways, and, turning to William, she added, "We need you. Be careful." Then she climbed on the wagon and with a flick of the reins headed the team back home.

He was stopped only once. Runaway slaves, he told them. He was returning them to their master who was hunting all over the state for them. He showed them a name and address in Columbus that he had scrawled on a piece of paper.

"Got this off the reward poster. The money will come in handy right now. I'm thinking they'll get their reward too," he added while nodding his head at the runaways. The men's laughter had sickened him for he knew they were enjoying the thought of the whipping awaiting the runaways.

Outside Columbus he concealed the runaways in a pile of brush. With determination, he made discreet inquiries in Columbus until he located someone willing to assist them from here. He was relieved and exhausted as he watched them continue their journey north.

Thus was his entrance into the Underground Railroad. Together, he and Nora assisted many others in their flight to freedom. He built a special bottom on his wagon to conceal his cargo, for the one bold journey had been enough adventure for him. They lost not one precious soul.

William did not live to see the Civil War tear the country apart, nor to see slavery abolished. He had been on another adventure; he and the boys had been hunting, when his horse stepped into a hole and threw him, breaking his neck in the fall. When Nora finally got over her grief, she preferred to think of him as merely off on a different adventure.

Inquisitive Nattie

It is now late in the Nineteenth Century, and this planet had entered a cycle that presented unique conditions and opportunities. In a desire to continue their growth, many personalities took advantage of this opportunity by rapidly re-entering the physical realm. As did this personality.

The beauty of life is that nothing is ever wasted or for naught. Everything you have done, are doing now and will be doing serves a purpose. Unfortunately, our unexamined expectations prevent us from realizing that fact.

Self-realization of what is contained within an individual is what motivates the personality to develop it. The motivation produced by your own self-induced realization that you are a very loving, spiritual and compassionate personality is 1,000 times greater than anything anyone can tell you. You gain that realization by acknowledging, examining and expressing your emotions. No other person can do that for you. Only you can. Once you start doing that, you become an example or mirror that allows others to see what is contained within them and what can happen if they develop that potential. In other words, our actions have a far greater impact upon people than our words.

"Nattie, say hello to your Grandpa and Uncle Ted," the girl's mother was instructing her.

She stood very close to Mama, arms wrapped around her legs, looking up at these two strange men. They were so big they frightened her.

"Come here, girl, let's have a look at you," one of the men encouraged as he knelt down and held out his arms to her.

Her Mama reached down and pushed her toward the man, "Go on, Nattie; let your Grandpa have a look at you."

Mama had told her they were coming, to help with the auction and to move them back to Grandpa's farm. It was far away, she said, a place called Illinois. She cried when Mama told her they would be moving. Then Mama started crying too. Mama has been so sad since Papa died–she cried a lot.

Nattie didn't know what an auction was, or that so many folks would be coming to their place. Mama just sat in her chair, watching and rocking away. Grandpa picked Nattie up and swung her up on his shoulders, so she could see, he said.

"Grandpa, them folks is taking our belongings. Make them stop!" Nattie cried out

He set her back down, took her by the hand and walked with her to his wagon. There, he lifted her onto the seat, climbed up and sat beside her.

"Nattie, you got to be a big girl now. Your Pa's died, he ain't here to plow and do the chores no more. You and your Ma are coming with me, your Grandma's back home waiting. Me and your Uncle Ted brung the wagon to haul you back. But we can't take everything. That's why all these people are here. They're giving your Ma money for the stuff they're buying. When the selling's done, we'll load the wagon and be off."

She sat quietly, looking around and thinking about what Grandpa had just told her. So they weren't just taking the stuff, they were paying good money for it. Still, she didn't feel any better. Somewhere inside it hurt, not like she was sick or broken; it was more like a great big ache inside–a big hollow.

Even the baby pigs were sold; they were all squealing like they didn't want to go. She felt like squealing too, but then Mama would start crying again, so she went looking for the kitty instead. Kitty was hiding in the barn; Nattie called and called until she heard her faint mew. Scooping her up into her arms, she rubbed her nose in Kitty's warm fur and carried her to the pile of straw in the corner. There she laid down, and as she cradled Kitty close to her she couldn't hold back her sadness any longer. She cried and cried; her heart was breaking and she feared there'd be no fixing it.

After the auction was over they finally noticed that she was missing, and Uncle Ted found her there in the barn where she had fallen asleep. Mama said she had called and called, she'd been so worried, Nattie must promise not to do that again. She mustn't scare folks like that. Uncle Ted wasn't mad or anything; he was really nice, and he asked her if she wanted to bring Kitty with her to Grandpa's place. She would have to watch her 'cause if she ran away they couldn't stop to hunt for her.

But Uncle Ted made sure Kitty couldn't run away. He found some old pieces of board and fashioned a little cage for her. Then he cut up some rope and tied one piece around her neck, and, handing the longer piece to Nattie he told her to use it to tie Kitty to the wagon whenever they stopped to rest.

It was a long way to Grandpa's farm. Mama didn't say much the whole time; she sometimes smiled, but it wasn't a glad smile 'cause her eyes were still sad. Mostly, Uncle Ted did the talking. He told Nattie all about Grandpa's farm–it was a big place, he said—lots of cows, chickens and pigs. Yes, he replied when she asked, they had baby pigs. Best of all, there was a little colt that could use the tending of a little girl.

They traveled and traveled. Nattie feared they might run out of road, they went so far. She was back in the wagon talking to Kitty when Uncle Ted called her up between him and Grandpa. Pointing ahead, he told her to look.

"There it is! Won't be long, and your Grandma will be standing on the porch, waiting for you."

Grandma was big and warm. Uncle Ted was right. Grandma was standing on the porch, her hand above her eyes so the sun didn't blind her. When they drove into the farmyard, her face broke out into the biggest smile Nattie had ever seen. Wiping her hands on her apron, she hurried down the porch steps and came to greet them. Before Nattie's feet had even touched the ground, she felt her Grandma's arms around her, rocking her from side to side while she covered her face with kisses. Nattie grew warm inside; it felt good.

It was a big house. Downstairs was a huge kitchen, the parlor and a bedroom where Nattie learned Grandpa and Grandma slept. Upstairs there were three more bedrooms—Uncle Ted slept in one of them. Mama had the same room she had before she married Pa, and there was a small bedroom for Nattie. Grandma had it all ready for her, with a pretty patch quilt on the bed and lacy curtains on the window. When she asked if Kitty could sleep there too, Grandma didn't seem to take much to the notion. But Uncle Ted put his arm around her and told her if she wanted to share her room with Kitty, well that would be just fine with them.

They ate supper, and afterwards, while Mama and Grandma cleaned up, Uncle Ted took Nattie to see the new colt. He was so funny-looking; his legs were way too long. But he came right up and nuzzled her like he needed her to be his friend. He didn't have a name yet, Uncle Ted told her. She would have to think of what to call him. Nattie didn't know any good names for a horse – she would have to think on that one for a while.

After she washed up, Mama took her up to her room. They knelt in front of Nattie's bed while she said her prayers. Mama had always listened to her prayers; she said it was their time together with the Lord. Nattie wasn't sure what that meant, but she liked having Mama there with her. Then Mama tucked her into her bed, drawing the quilt up to her chin before picking up the lamp and leaving her there alone with Kitty in the dark.

She lay there listening to the sounds. She could hear the others going to their rooms. Then it got really quiet. And Nattie got scared. She crept out from under the quilt, picked up Kitty, and slowly felt her way to the door. It was so dark.

"Mama," she whispered, "Mama, where are you?"

Mama heard her, came to her asking what was the matter. Nattie asked if she could sleep with her. Just for tonight, she promised. Mama let her crawl into her bed and Nattie cuddled up close to her. Kitty curled up on the pillow next to her neck and began to purr. It wasn't scary here; if Kitty was purring everything was okay.

Grandma and Mama were always busy. It seemed that if there wasn't washing to do, there was cleaning needing doing. Or else, they were cooking or out in the garden chopping at those old weeds with the hoe. Grandpa never said much. She tried asking him some questions but didn't get much of a reply—mostly, just a grunt or something. But Uncle Ted was different—when she asked him a question he would stop what he was doing and talk to her. Nattie liked this, and decided that he was her special uncle.

"Did you think up a name for the colt, Nattie?"

"Yep, Broom," she replied.

He laughed and asked her how on earth she ever thought that one up. "His legs looked like broomsticks," she told him. So she decided his name should be Broom.

"Uncle Ted, how is it you ain't married like the rest of my uncles?"

She had been wondering about this for some time. Now the words just kind of popped out of her mouth. She thought maybe she shouldn't have been so nosy 'cause he stood there looking at her with a funny look on his face. Then he sighed and told her he figured it was 'cause no good woman would have him.

"Well, I will. I'll grow into a good woman and I'll marry you."

He laughed and said he was certain she would grow to be a good woman, there was no doubting it. The way he said it she could tell he wasn't just funning her, he meant it. Truly!

The following year she started school. The country schoolhouse wasn't all that far from the farm; when the weather was good she walked, carrying her pail of lunch. Otherwise, Grandpa or Uncle Ted would drive her there in the wagon and pick her up again later. It seemed like a lot of fussing though, all that hitching and unhitching just so she could make it to school.

Nattie learned her numbers and letters. Each night she would sit at the kitchen table while Mama and Grandma were cleaning up, practicing on the slate board Uncle Ted had bought her. Mama would come over to see what she was doing and stroke her hair while she watched. It made Nattie feel like everything was better; Mama even laughed again once in a while. Nattie knew she still missed Papa 'cause Nattie missed him too. Grandma had explained to her once that the Lord had called him home. Made no sense to her 'cause he had a fine home here with them. But, Grandma said that was the way it was and they had to make the best of it. So, Nattie tried really hard to be good.

Her school years passed quickly. Graduation from the eighth grade was cause for celebration and they had a party at the schoolhouse. Mama baked a special cake and carried it over; another mother made ice cream; there was lemonade and lots of cookies. Nattie felt proud and sad, all at the same time. She was proud of herself for doing well in school, for passing all her tests. But it made her sad to think that she would not be seeing her friends every day. Now she would have to saddle Broom and ride over whenever she wanted to talk to one of them.

That Sunday, all her aunts, uncles, cousins as well as many friends came to the farm, and even her teacher was there. Mama and Grandma had been cooking and baking all day Saturday, rising early again Sunday to do some more. Grandpa and Uncle Ted carried sawhorses and boards to make tables out in the yard—after they were covered with tablecloths, they looked like real tables. All the chairs they could find were carried outside; still, there wouldn't have been enough except that Uncle Ted had made two long benches.

Everyone made such a fuss over Nattie. It didn't feel comfortable to have so much fuss made over her. She even got presents, lots of them: books to read, writing paper, her own Bible, too many to remember them all.

Uncle Ted looked strange. He wore a white shirt under his new bib overalls, and his hair was all slicked down and shiny.

"Look at your uncle, Nattie. He's sweet on your school teacher," her Mama said as she wrapped her arm around the girl's shoulders.

Nattie remembered all the times he had come to pick her up at school. She hadn't paid it much mind that he always had a word or two with the teacher, just thought he was being friendly, like always. She liked her well enough, but she couldn't understand why she felt like she did about her and Uncle Ted. It took some of the fun out of the day for her.

That night Nattie went to Mama's bedroom and asked her if they could talk a bit. The two of them sat on the bed while Nattie tried to sort through her feelings.

"I know I should be happy for Uncle Ted, he's always been so kind to me. And I do like Miss Appleton, I really do! So, why do I feel so miserable inside?"

"Oh, Nattie, you're just feeling a little jealous right now. You've always had Uncle Ted all to yourself. Now all of a sudden you find out you have to share him with someone else. It's just going to take some getting used to, you'll see."

Nattie hoped she was right; she didn't like the way she felt.

It did get easier. Miss Appleton was at the farm a lot that summer, helping Mama and Grandma with their chores. It was kind of funny, Nattie realized. When she was going to school, she just looked at Miss Appleton as her teacher, she never thought of her in any other way. Now at the farm, Nattie started to see her differently–it was almost like she was a different person all of a sudden. She laughed and teased Uncle Ted; sometimes he got all red in the face. By harvest time Nattie realized she liked Miss Appleton better now that she was just like everybody else.

Harvest meant everyone worked. Grandpa hired extra men to help bring in the crops, so there was a lot more cooking to be done. Nattie washed dishes till she thought her hands were going to shrivel up from being in so much dishwater. Then they had to can what was left in the garden and pick the rest for storing.

"Nattie, what's wrong?"

"Nothing, Mama. Just a stitch in my side."

But at suppertime her side still hurt. She hardly ate anything, so Mama knew something was wrong. They figured she'd been overdoing it with all the working, and they made her go up to bed.

Mama came in looking very concerned, and asked Nattie if she'd had the pain before. It had been there off and on for a day or two Nattie realized; she'd ignored it, thinking it would go away.

During the night it got really bad. Mama heard her crying and came in and lay down beside her.

"Mama, it hurts something fierce."

"I'm waking Pa. Somebody's gotta go get the doctor!"

It just hurt so badly. It must be serious if Mama thinks she needs the doctor. Maybe the doctor can stop it from hurting.

When Mama returned, Grandma was with her and came over to feel her forehead.

"The girl's running a fever!"

When Nattie opened her eyes again, Grandma had a pan of water sitting there and was wiping her brow with a damp cloth. She felt really odd and her eyes didn't always see so well. At times everything sounded far away, like it was in another room.

The doctor was there, bending over her. He was talking to Mama but Nattie couldn't make out what he was saying–just a word here and there.

Mama was sitting beside her bed in her rocker, asleep. Someone must have carried it up here for her.

"Oh, Mama, look. There's Papa. He's come back. Oh, Papa, I'm so happy to see you again. I don't hurt any more, Mama."

Summary

The life experiences of Angel Lyn are examinations in faith, emotions, and trusting our internal guidance system. Meditation allows you to experience the absence of thought and experience the insightfulness of consciousness. Paying attention to your emotions and expressing them provides additional insight. Allowing yourself to experience both develops faith and trust in "self".

Every incarnation serves a purpose, even those which appear obscure and without complication. The life experience of Pastor John was certainly filled with complications; not only was there the emigration to America but also the added complication of the Massachusetts Bay Colony. The life experience of Pastor John is a good example of acting on faith – an inner knowing and trust in yourself and your guidance system.

You're probably very familiar with the word "faith" for you surely have heard it expressed enough. Therefore I ask you, what is faith? What prompts one to respond as Pastor John did? Is faith a trust placed in something outside of you, or is it a trust in our own inner guidance?

Do you see how familiarity with the words you use causes thought to think and believe it understands them? Yet, when you ask, "What does that word mean?" you realize that you really don't understand it.

God does not ask you to blindly obey but rather that you come to your own understanding in your own way. Thus, the purpose of this incarnation was for Pastor John to experience listening and trusting in him while encouraging others to do likewise. It was another examination of the human spirit – the innate desire to live in freedom regardless of the difficulties encountered.

Every journey begins with the first step, and then another and another. So, too, our journey of evolution begins with one incarnation, and then another and then another. Each step brings you closer to realizing through experience that you are the personality.

This personality was very spiritual and had a deep love for God. Many emotions washed over me as I recorded what Pastor John experienced. Firstly I became aware of the love this personality felt for his people. Next, the pride he and his people experienced as they faced one difficulty after another without wavering in their faith. Finally, I became aware that he had little concern for "self"; most of his thoughts were for others. Although his wife and children were important to him, he did not cherish them beyond all others.

This personality chose quite an experience for itself as William. Although it wasn't an easy one, you'll have to agree that it was one heck of an adventure. It is an excellent example of the fact that everything isn't as it appears to be. Only Nana could see that William wasn't the "black sheep" of the family, as everyone had thought – she was able to recognize something in William that others couldn't because it was contained within her. Having experienced it herself, she understood that it would emerge when the proper set of circumstances came about. William proved her correct.

William's true personality didn't emerge until he got away from the dominance of his father and the shadow of his brother. In Ohio William found the ready acceptance of friendly people; people who expected nothing more than a helping hand when needed. In the classroom he discovered enjoyment when he allowed himself to experience himself instead of attempting to fashion himself after other educators. Thus his students respected him and regarded him as their friend.

To the dismay of his father, William continued to follow his inner guidance. It didn't matter to him that he didn't receive any recognition for marching to a different drummer. However, in time, even his father came around and signed the property in Ohio over to him. Curiosity and the ability to view things as *possibilities* caused William to explore and examine those possibilities rather than pursue the accumulation of material wealth like his father and brother. Consequently, his life was very fulfilling and rewarding. In fact, without any effort on his part, he experienced material wealth as well. What more could anyone ask for?

The discovery of the runaways tested William's compassion. Assisting runaway slaves endangered his life and those of his family. Once again William heeded that still voice inside himself: he knew he had to help them. The boldness of his plan and the execution of it contributed to his success in helping the slaves. Who would expect a person to lead runaways down the road in broad daylight if he was helping them to escape?

William never realized what a kind and generous person he was.

Nana knew. Nora and his children knew, along with countless slaves who found their way to freedom with his assistance.

Writing this, I became aware of a host of personalities wishing to convey their thanks. There never was enough time for them to tell him what his kindness meant to them. They were always hurried along and told to be quiet. Many repeatedly said, "No time for thank you, just hurry before someone comes."

I hope you wondered about or questioned the purpose of the incarnation as Nattie, especially when she died so young. Emotions are the language of the personality. You're unable to see that if you ignore, rationalize, or deny them, and the same applies if you don't express them. Acknowledging and expressing emotions is the common element in all three incarnations. Contained within each are several examples of when this personality acknowledged and expressed its emotions, especially as Nattie. Although Pastor John was very loving, he frequently did not acknowledge his emotions, and concealed the ones he did acknowledge. That's because he paid more attention to thought or the mind than to the heart.

What the personality experienced as Nattie confirmed what it had discovered as William and strengthened its desire to examine business. Finally, the incarnation as Nattie facilitated the personality in changing its avenue of service. Therefore, you will not be surprised to learn that today this personality is self-employed. Examining your life allows you to see that everything you experience provides the guidance you seek. Furthermore, everything you experienced so far has prepared you for what you will experience today.

Take a few minutes to reflect upon the last time you had unwavering faith in yourself. Carefully examine how that faith made you feel and act. After you have done that, examine the last time you really doubted yourself. Examine how you felt and acted. Do you see what factors caused you to have faith in yourself and what factors prevented you from having faith in yourself?

Chapter 6: Angel Mara

Earl William

This incarnation takes place in England during the late sixteenth and early seventeenth centuries. The Scottish King James VI was about to succeed Elizabeth I and become James I of England.

Transitions in government are always difficult, and England's future rested on the wisdom of its parliament. Will those who have the "right stuff" step up to the plate and assume their responsibilities? Or, will those who have the intelligence to serve in Parliament neglect their responsibilities?

John Bodley, a politically well-connected Protestant merchant, had recently married Ann Ball, a widow of considerable fortune. The following story is about an unexpected event that happened to this family and how John Bodley dealt with it. It's an unusual life experience that is covered in secrecy.

His mother was calling him again. She would soon find him and bring him back inside. It was not that William did not want her to find him - he was not a disobedient child; he merely wished to remain on the balcony watching for another horse-drawn carriage to go by. William was unused to so much activity; they had arrived only the previous day from their home in the country.

Oh, there she was, calling him again. He'd better go to her before she became angry. William does not it like when she is angry – like she was with Father this morning. That was when he came out here, so he would not hear the angry words. And there was so much to look at out here, so much to see, like that carriage coming near.

"Here, Mother. Here I am," he called to her as he reentered the room.

"William, I have been looking all over for you. You must learn to come when you are called. We have discussed this before! You are too big to play these tricks on me," his mother admonished him. Continuing, she told him, "Now we must hurry that you are properly attired for soon we will have a caller."

He did not enjoy being "properly attired" – it felt most uncomfortable. It was frilly and too tight. In his play clothing he could run and jump but being "properly attired" one could do nothing but sit quietly. That was no fun at all.

"There, what a fine little man you are now," his mother said to him as she admired how he looked. "See how handsome you are," she said as they stood looking in mirror.

Together they walked to the main parlor. As they entered he saw his father already speaking with someone. They did not sound much like friends for they appeared to be arguing, but when the stranger saw him and his mother he smiled and rose from his chair.

"This is good of you," the man said to his mother before turning to William. The man stared so long at him that William became uncomfortable, shifting his balance from one foot to the other. The shoes hurt his feet. He wished he could either sit in one of the chairs or return to his room and remove the shoes.

"Enough," said his father. "This is all we agreed to. Take the boy back to his room."

His father always had a stern look on his face but now even more so, William observed. Confused, he looked to his mother, but she just smiled at him. She had been standing with her hand on his shoulder, now she gently squeezed it and guided him out of the room.

William missed his brother, but not his sister. She was still just a baby. If his brother, James, was here also they could eat their supper together, but now he had to eat with the cook. She had been told to put him to bed before his mother and father left for the evening. His mother was so beautiful in her new gown, and William had told her she must be the most beautiful mother in all of England. His father was most handsome in his evening dress. William had started to tell him so but stopped when his father scolded him for touching his mother's gown.

Cook listened to him say his prayers before tucking him into his bed. Waiting till he was certain she had returned to the kitchen, he rose and went to the balcony. Horse-drawn carriages were passing by, so many of them! He wondered whether everybody in England would be attending the ball tonight. He watched until the evening chill began to settle upon him, when he returned to his bed and was quickly asleep.

They remained in London two more days. His mother had taken him for a long carriage ride, stopping all too frequently at shops that took her fancy. He much preferred to ride on, to see the many buildings and fine houses. When they returned to the house William had helped carry in the many packages. Then she had kissed him and told him she must be off. She had one more engagement that day. His father was already gone so he would dine again with cook.

He was happy to be returning home. The city was most interesting if one could but see it all, but he had not even been allowed to play outdoors during their stay; instead he remained couped up indoors, looking out. At home he and his brother were often outdoors. He loved to visit the stables, especially if there was a newborn foal.

The journey home was pleasant. His father remained in London and his mother told him stories to pass the time, pausing only to draw his interest to something along the way. His brother and his aunt were waiting to greet them as the carriage drew up in front of the house. Locating the toy horse he had bought his brother as a gift, William jumped down and handed it to him. As they were about to run off and play, his mother asked him if he did not wish to tell his aunt everything he had seen and done. Reluctantly the two boys followed to spend the remainder of the afternoon indoors.

The two boys were tucked into bed and should have been sleeping, but a loud noise as though something had crashed to the floor awakened William. Now as he lay listening, he heard his father talking loudly to his mother. He could not hear what they were saying, but he could tell his father was very angry. Creeping slowly out from under his covers, he tiptoed over and opened the door just enough so he could hear their words.

"This is still my house and you will do as I say. You're still no better than a whore. I never should have married you," his father roared.

"Yes, and what would you have done then? What if I were to tell him the truth, what a fine gentleman he married me off to?" was his mother's reply.

William cringed as he heard his father slap his mother, and moments later he could hear her sobbing as she ran past his room. Quietly he closed the door, fearful his father may discover him. He remained there until the house had grown silent, then he crossed to his brother's bed and crawled in beside him.

It was their aunt who breakfasted with them, telling them their mother was not feeling well and would rise later. They should eat and then go out to play, perhaps visit the stables, she suggested cheerfully.

William's father summoned him to the library. Seated at his desk he did not look so stern, and William relaxed when told to sit down. If he were to receive a scolding, his father would have left him standing. He tried not to squirm; he sat like a gentleman, as his mother called it, waiting for his father to speak. His father sat back, his cheek cupped in one hand, looking at William.

"I had not realized how you have grown, William. You are no longer a little boy to be coddled all day by women. It is time for you to put aside your playthings that you may prepare yourself for manhood. Arrangements have been made for you to attend a fine school. There you will learn everything a man needs to know. What you do with it is up to you."

William did not understand. There was no school nearby, so how then was he to attend? He was about to ask when his father straightened in his chair, his face once again stern.

"You leave in the morning. Your mother will pack and have you ready. Be off now. You're excused," he said abruptly without looking at William.

William returned to his room to find his mother already packing his clothing. It was true then, he was being sent away!

"No, Mummy. I do not want to go away. I'll be good, I promise! Please, don't send me away," he begged.

"Oh, William, I know you are a good boy, so does your father. But as boys grow there comes that time when they must learn things, such as how to read and how to write. You see all the books in your father's library – how will you ever read them if you don't go to school?" she asked him.

"What about James? Is he going, too?"

"No, he is still too young, he must remain here." Before William could say anything further his mother continued, "William, I love you very dearly. It is not easy for a mother when her children grow and leave her. I ask that you do not make this any more difficult for me for my heart could not bear it."

Her voice softened as she went on to reassure him, "It is a very fine school, William. This is a wonderful opportunity for you. There will be other boys the same age as you and you will learn all manner of wonderful things. Besides, you will be returning home every holiday and every summer. Then you can tell James everything you have learned!"

William tried very hard to be brave. The carriage was waiting outside as he said good-by to James and Sarah, his sister. His aunt kissed his cheek as she hugged her farewell. When his mother knelt before him and took him in her arms he could hold back his tears no longer. Laying his head on her shoulder, he wept as she stroked his hair and told him to be her brave little man.

"Come now, William," his father said, clearing his throat. "It is time to be off."

The fine school was Eton. Although he did not excel in his studies, William adjusted to his new surroundings more rapidly than some of the other new boys. He most enjoyed numbers, and each holiday worked with James until he could finally remember them. Spelling he liked least of all and it was the most difficult for him of all his lessons. Returning home every summer, he discovered he was always ready for school to begin again.

And now his years here at Eton were at an end. His room stood bare of all his possessions, and he observed that it appeared lonely without books and clothing as he paused for one final look before turning and walking out to the waiting carriage.

"William? Yes, it is William, is it not?" a man said to him

"Yes, my name is William," he replied. "But I do not believe I know you, sir."

"No, you would not remember. You were just a boy when last we met, and that was a long time ago. You have grown into a fine young man. Your parents must be most proud of you," the man answered without volunteering his name. "Well, I must hurry off. I'm already late for my engagement. So good to see you again."

As the carriage pulled away William looked back. The man was standing, watching them drive off. Odd, perhaps a bit daft in the head, he thought as the school of his childhood disappeared from view.

"Mother, why has James never been sent away to school? It is unfair that he still remains here with that tutor Father retained. Just today I posed a simple question that James was unable to answer. He deserves better teachers, Mother," William was arguing.

"This is none of your concern, William. James does not have your mind; it's not easy for him to learn. Better he remain here than be sent off somewhere to be ridiculed."

"I know it is not easy for him. It was not easy for me either. I am saying that it is unfair that I'm going off to Oxford and James is to remain here with a simpleton for a teacher. Perhaps if I did not attend Oxford, those moneys could hire a better tutor for him."

"I already told you this is none of your concern. I know you love your brother and your concern is admirable. However, you will attend Oxford! That is already decided and is not open for debate. Your brother is receiving the best education he is capable of. Please, William, do not make your brother feel worse than he already does by continuing to raise this issue. James has accepted that he is not as bright as you are, why can't you?"

She was right, he knew. James often asked him questions such as a child would ask. Within him, he had known for a long time that James had difficulty learning and remembering, but he had hoped it was something he would outgrow. The time they now spent together was no longer carefree as it had been when they were boys. They no longer shared the same interests, and it was more than years that now separated them. With a heavy sigh, William gave his mother the promise she was seeking. He would accept that his only brother should remain here, surrounded by those who loved him and accepted him as he was.

During his years at Oxford, he formed lasting friendships. His holidays were often spent as a guest at one of their homes. He was hesitant to invite any of them to holiday at his home, uncertain how they may respond to James. But, none seemed to notice, or if they did they were too well-mannered to comment.

His interests at first were the world of finance as his ease with numbers developed into a fascination, and he thought this might be a profession he would consider. But he discovered that he enjoyed a good debate more than anything else. Among his friends, a loud groan would arise whenever he challenged one of their statements, and challenge he did, every chance he had! When one of them suggested he sounded more like a member of parliament than a friend, he responded there was no reason why he could not be both.

Graduation is a most solemn occasion, they were told. It was time for them to go forth with their education and continue to make England the finest nation in the world, the speaker continued. William could see his mother and father sitting and listening attentively to the speaker. They do make a handsome pair, he suddenly realized. How is it that I never noticed before, he wondered, before realizing that he now was mature enough to see them as real people.

At last the speeches were done. They all stood and began to walk up to receive their degrees, and everyone applauded most politely as each was congratulated. Finally, it was William's turn. As he approached the podium he now could see the man waiting to hand him his degree. Aware that he was one of Oxford's benefactors, William had paid no attention to his name. But, somehow this man appeared familiar to William, although he could not recall where they had met before.

"Congratulations, William," the man said as he shook his hand.

Thanking him, William had to move on. There were others behind him waiting. He decided he would ask who the man was later. Or perhaps he may even have the opportunity to speak with him at the conclusion of the ceremonies.

Then it was over and the proud parents crowded around their sons, anxious to add their congratulations. Mothers were wiping tears while sweethearts looked on. William had invited Mary Elizabeth to attend so that he may properly introduce her to his parents. The sister of one of his friends, he had met her at a party. They planned to marry, but had told no one. He wanted time to create the life they would share rather than marry in haste and end up with the life given to him. She charmed his parents. It became apparent that his mother approved when she inquired if Mary Elizabeth would perhaps enjoy a summer in the country. His parents had remained protective of James, shielding him from every possibility of insult or ridicule. For his mother to now invite a stranger into her home indicated the trust she already felt for this young woman.

William floundered about a bit as he searched for a profession he felt was of merit. When his father died, the title of Earl passed on to William along with the right to a seat in parliament. Why his father had never accepted his responsibility as a noble, William could not understand. To him, a man had no right to a title that he did not uphold. His future now lay clear to him; he had always welcomed debate. This was what he had been preparing for all these years.

He claimed his seat in parliament and a month later he and Mary Elizabeth married. They had spent two weeks in the country with his mother and James and Sarah, now also married, had joined them for a week along with her husband and infant son. The small child was the center of attention and James, especially, enjoyed playing with his nephew. He retained the innocence of a child yet himself. When it was time for their departure, James had bowed before Mary Elizabeth before telling her how happy she made him feel.

When they returned to London, William found a small package had arrived for him. Studying the handwriting as he walked into his library, he was curious who would be sending him something now. It was not addressed to both of them, so it was not a belated wedding gift. Nor was it his birthday or any other such holiday. The package remained unopened for several minutes as he pondered the sender. With the mystery still remaining, he finally unwrapped the package and found an ornate box containing a handsome gold pocket watch along with a letter.

Dear William,

It was been my good fortune to witness you claim your seat in parliament. How rare it is to discover concern for the affairs of state awakened in one still young. England can only be the better for it as long as men such as yourself come forth.

Please accept this small gift as a token of my admiration.

Sir Thomas Bodley

How strange, he thought, as he examined the watch. The case was exquisitely carved with his initials. Inside was the inscription, "May you walk forever in righteousness." The sender's name was one he had heard before, and now he struggled to remember who he was. When Mary Elizabeth joined him, she could offer no assistance, adding only that she was proud to be married to a man so apparently admired by others.

His mother died shortly thereafter. She had not been ill; the doctor said her heart failed her suddenly. The care of James now became his concern. It would be unfair to move him to the city this late in his life, so instead William and his wife moved out to the country. William commuted to London as needed, remaining in their townhouse during his stay. It was there one night that he recalled the name 'Sir Thomas Bodley.' He was sitting in front of the fireplace, drinking a brandy before retiring. As he relaxed, he fingered the watch as he had done many times before when deep in thought. Suddenly, he recalled the name and knew the man it belonged to. This was the man who had spoken when he graduated from Oxford. The man who he thought looked familiar and then forgot about once the celebrations began. Yes, it was on his last day at Oxford, that's where he'd heard the name. And, he had also seen him on his last day at Eton, he now recalled. A strange man indeed, and, remembering that he had thought perhaps he might be daft, William laughed – for this was a highly respected man. Not daft, but strange. A pity I did not remember before, he thought, for the man had died shortly after his mother, too late to thank him now. Sighing, William rose and went to bed.

John Quincy Adams

Imagine if you could return to colonial New England during the days of the American Revolution. What would it be like to hear cannon fire all around you and observe the smoke from countless fires? As you looked upon the horizon, could you tell which towns were burning? How concerned would you be for your safety and that of your family? Today we realize what an historical event it was; however, the colonists did not know they were witnessing history in the making. They were merely people, living their daily lives with the same concerns we each still have today. They were concerned with the safety of their homes and loved ones as well everyday matters such as the taxes they were paying.

The life experience of this angel is an examination of responsibility and independence. It also reveals how the daily actions of parents influence and impact their children. Regardless of how difficult times were, the daily display of love, responsibility and wisdom by John's parents enabled this angel to become a very responsible and independent man.

John's father was away again. He had been admonished to heed his mother, tend to his lessons and remember his chores until his father returned. Although he was only seven years of age, he took the words of his father most seriously.

"These are serious times we find ourselves in, John. You are a child who cannot enjoy the luxury of a childhood for in my absence it falls to you to care for your mother and sister," his father had told him just before he left.

First there had been the British attacks on Lexington and Concord. When word reached their village, young John had not slept easy. He wondered how he would protect his mother and sister, Abigail, if the British should attack. For days, he kept watch, expecting to see the Redcoats marching into town at any moment.

It was a great relief for him when his father returned, although his relief was short-lived. His father would not be long at home, he learned; it would soon be necessary for him to leave again. Just as many others were doing, his return was to assure himself that his family was safe and that their needs were met. John prayed his father would remain; he wanted his father home where he could go to him whenever he needed. But it was not to be. The family prayed together for a righteous resolution to the differences between England and America as his father prepared for departure.

"Mother, Mother, come quickly," John cried as he ran to their home.

Hearing the urgency in her son's cry, his mother came outside to investigate. He pointed at the smoke that was visible in the sky. Fires were an ever-present danger – if this was a forest fire it could threaten the safety of their home.

They stood watching along with others who had observed the smoke. "The fire is to the north. I pray dear God that it is not Boston burning," someone said. From another, "Could it be the British? Did they set the fire to burn out the rebellion?" Listening to the people, John moved closer to his mother, needing to feel the nearness of her.

Then from a distance they could hear the sound of cannon fire. A silence fell over those assembled as they stood watching as though they may be able to see something, anything that could tell them what was happening. Looking up into his mother's face, John asked, "What is that sound, Mother?"

"That is the sound of war, son; those are cannons firing that we now hear."

"Who is firing the cannons?" he asked, "Are they our cannons or England's?"

She was unable to answer. No one knew; they could only pray it would soon be over. With a sigh, she finally told John it was time for them to go inside, time for supper and to finish up his chores for the day.

He had little appetite. Sitting at the table, he listened. Several times his mother reminded him of his plate of food and he would eat a bit more before stopping again to listen. As darkness fell, all became quiet; it seemed the whole world was silent, he thought, for not even the call of a night bird could be heard.

The next day everyone was talking about the cannon fire. Rumor after rumor spread through the town, each one worse than the one before. It was several days before they learned what had really transpired – that the Colonists and the British had fought at Breed's Hill and Bunker Hill. The Colonial forces would have been victorious if they had they not run out of ammunition.

Elated, John ran to tell his mother. He expected she would be as happy as the others; he did not expect the somber face he now saw. When he asked her why she looked so sad, she replied, "Men have died, brave men who left their families to fight for what they believed to be right. Whether they were one of ours or one of King George's no longer makes any difference for now they answer only to God." Suddenly he thought of his own father and was thankful he was safe.

They continued to hear rumors that the war was continuing, with battles fought daily. Each time his father returned, they were able to learn the truth. Patiently he would sit and answer each of John's questions until none remained. Then father and son would go for a walk about the village, where his father would encounter more questions. Listening intently, John was astonished to learn there were those who remained loyal to England. When he asked his father about that he was told they must be forgiven for they did not know any better.

The onset of winter slowed the rumors; most were more concerned with the weather and whether or not they needed to chop more wood. His father was home, the smells of his mother's preparations for Christmas filled the house, and John was happy. Forgotten were his worries of the summer – it was time to celebrate the birth of Jesus. He loved the story of his birth; with everyone gathered round, his father would open the Bible, look to make sure everyone was paying attention, and then begin to read. John knew it by heart, he could have recited it word for word, but he loved to listen to his father read. Now here, safe in their home, his heart felt as though it was filled to the top and ready to overflow.

Spring arrived and again the rumors began. In the absence of his father, John gathered every scrap of information he could find so that he could learn enough to be forewarned of any impending threat to their safety. As did so many others, he often paused and listened to hear if the sound of cannon fire could be heard.

As he went about his chores, John found a young bird with a broken wing by the woodpile. It must have fallen when it was learning to fly, he thought, as he carried it in to show his mother. He asked her if he should find the nest and place the bird safely there so that its mother may care for it.

"How will you know which nest to place it in? It may be far away from its nest. If the mother was teaching her young to fly it is time for them to care for themselves – she will no longer feed them."

"But, Mother, the bird is hurt," he replied.

"Yes, John, it is. But the mother does not know that, she only knows it is time for them to be on their own – she cannot take care of them forever."

"Is that like us, Mother?" he asked.

"What are you asking, John?"

"We are far from England just like the bird may be far from its nest. And Father said it was necessary for us to care for ourselves while King George kept his armies busy elsewhere. Now he wants us to obey his orders again."

"When this bird is whole again and is able to fly and find its own food, would that bird wish to return to the nest of its mother and do its mother's bidding? Or would that bird build its own nest that it may decide what is best?"

John stood pondering. He loved his mother and was not over-anxious to be away from her. "But, when I am grown to manhood I must make my own way. As Father does," he realized.

"We will care for the bird while it mends. There is grain in the shed and worms and bugs are plenty that you may bring it. It will fly away when it is ready, when it no longer needs us."

His mother was truly wise, he thought, for she had been correct. Daily he searched for worms and bugs under rocks and in the garden. The little bird quickly devoured everything John brought and he could see it growing stronger.

Then one morning the bird was gone. John had found several bugs that he thought would look tasty to his bird, but when he went to give them to the bird, it was no longer there.

As he sat to breakfast his mother asked him what was disturbing him so. His bird was gone, he told her.

"That was not your bird, John. It was only yours to care for until it grew strong enough to care for itself. You were most diligent in your duty, and the bird mended well in your care. Now it flies free because of you. You too are free; you no longer are responsible for it. Instead now you are free to do other things."

He knew in his heart his mother was correct. Times were he did not enjoy searching out the bugs; he would rather have busied himself elsewhere. He had not admitted it while the bird still needed him, but now the bird was gone. "It no longer needs me," he thought, "maybe it will return to build a nest here next year." He relished the thought of the bird nesting near his house where it knew it could be both safe and free.

John was already sleeping when his father returned, but the sounds of his parents talking awakened him. Too sleepy to rise, he rolled over and returned to sleep, happy in the knowledge that in the morning his father would be there to greet him.

The morning time found others seeking out his father; John was disappointed to discover his father already speaking with them when he arose. The times when he and his father were together were his favorite – times not shared with any others. Now it would be necessary for him to wait until these men left.

"John. Come here, son. Sit with us," his father called out when he saw the boy was up for the day.

The small parlor was filled with men all looking to his father. John wondered what had brought them to their house so early in the day. Was there more trouble to fret about? If so, he was most happy his father had returned because he would know best what to do.

"Gentlemen, I need first to speak to my son, to tell him of what has taken place," his father said before turning to John.

"Son, on this day our little colonies become a nation. It is our hope they will someday be marked as a great nation. Independence has been declared, independence from England. We must decide what is best for us – not some King who resides across the ocean, even though a righteous man he may be. He is not here nor we there. Remember this day, John. This is the day our nation declares itself free."

As his father spoke, John remembered the bird he had cared for. In his mind he saw it flying free, soaring to great heights. "Perhaps our colonies are like that bird," he thought, "far away from the nest of England we cared for ourselves, and now we are attempting to fly. How high," he wondered, "How high will we soar?"

The Revolutionary War followed and, during that period, John traveled twice with his father to Europe. He entered Harvard after returning from living in Hague, London and Paris, where he had furthered his education. He now knew several modern languages as well as Greek and Latin.

At thirty years of age, John married. He and his wife named their first son after the first president of the country, and their second son was given the same name as his grandfather, John, and the third was named Charles Francis. They were greatly saddened when their second child, their only daughter, died in infancy.

There are many who believe this man was forever overshadowed by his father and failed to achieve the same greatness. That is untrue. One needs only to examine what history has recorded to discover that John Quincy Adams was a man forever dedicated to his country. He made no attempt to walk in his father's footsteps; instead, he determined his own future. It was a future that would see him serving his nation till the day he died.

Joseph, Mountain Man

Shortly after the Louisiana Purchase in 1803, the US Government encouraged newspaper reporters to entice people to settle this vast and uncharted area any way they could. Some reporters painted unrealistic pictures of the fortunes settlers could obtain west of the Mississippi River. But the greatest motivation to move west came in 1849 when gold was discovered in California.

We cannot know or miss what we have never had. If one has never known anything other than loving relationships, how does one ever come to fully appreciate and value their worth? If we can understand that concept, we can see that the personality of this angel chose to experience the lack of what it experienced in the previous incarnation so that it could obtain a more complete appreciation of how these things assist in one's personal growth. This process enabled the personality to examine responsibility, independence and self-reliance from a different perspective.

His parents were among those who migrated to Colorado in search of land. After the discovery of gold there, word had spread of the vast prairies lying in wait of a plow to turn them into grain fields. The mining towns cropping up all across the state had need of provisions, and many felt it was a ready market for a farmer willing to break the land and raise cattle as well as grain for flour.

But life there proved to be hard. No one had told them that this land thirsted for water. In the Ohio Valley they had left behind, the land turned green and lush every year, blessed with rich soil and ample rainfalls. This land did not yield easily to the plow; the parched soil was hard and difficult to turn over. Once tilled, the winds sweeping across the plains would raise great clouds of dust, blinding both man and horse.

The boy's parents had been more fortunate than most; they had brought with them two cows and a bull. His father fancied the day would come that he would fulfill the hungry miners' appetite for beef. He was a visionary; unfortunately, he had not known that water would prove to be his greatest challenge.

Their first child had been stillborn during their trek to Colorado; the second lived only until its first year. By the time Josef was born, fearing he too would be taken from them, his parents dared not become overly fond of the boy. However, their strong German heritage had finally produced a child with a will to survive. It was as though the best their genes could summon went into this one child and thereafter could not be repeated. His mother never conceived again.

From the time he was old enough to leave his mother's side, Josef was his father's shadow. Trying to win some indication of approval, the boy imitated his movements, even walking as he did. His father was neither demonstrative nor talkative; both demanded time, and work filled his time.

Had there been a ready supply of water their herds would have grown and provided for this small family. However, each spring the waters would rush across their land and the grasses would turn green. Then under the heat of summer, the land would dry and the grasses turn brown. The cattle remained lean, and each year at least one was lost to thirst. Snow, visible on the mountains off in the distance, mocked them. His father would stare at the snow and shake his fist at the mountains, as though trying to dislodge it. Josef would look and see the promise of escape from another hot, arid summer.

In the spring of his fifteenth year Josef's father left to replenish their supplies, which had been depleted by winter. Josef wanted to ride with him; the opportunity to talk with others appealed to him. However his father stubbornly refused: the chores needed doing, so Josef must stay home.

It was a day's journey; even if the weather took a turn for the worse, his father should have returned within two days. Two days passed, then three and four, and still they anxiously waited. But he never returned, nor was any trace of him ever found.

His father's unexplained disappearance seemed to drain the life from Josef's mother, and every day she grew sadder. Josef tried to provide her some cheer, bringing her a newborn calf to admire or picking a wildflower he had come upon. Nothing revived the brightness of her eyes; she had lost her will to live. It really came as no surprise to him when he arose one morning and found she had quietly died in her sleep. He buried her there on the prairie that had been his parents' hope for the future.

The prairie had won. Josef had no desire to remain and continue to endure the same hardships. With the horse and wagon gone, he bundled up only what he could carry. The herd of cattle his father had dreamed of numbered only seven cows, a bull and one calf. On foot he herded them to town and sold them. Had they been fattened by the summer grasses they would have brought him more money, but he took what was offered, bought a horse and rode off.

He knew he had suffered his fill of the prairie – it was the high country that he was drawn to see. The snow that lay on the mountain peaks, even in the heat of summer, told him water was never a need, and that appealed to him most of all. He rode in leisure, free of all responsibilities except to provide for himself and his horse. He had no vision of his future, nor was he concerned. He felt he would ride till he found a place he wished to remain, and there he would stop and put down his stake.

For the next three years or so he wandered about the mountains. He tried his hand at panning gold, even though the gold rush was over. Quickly he determined it was a waste of his time and moved on. He became acquainted with many of the men who made the mountains their home, camping with them whenever their paths crossed. Although he enjoyed the encounters he was always ready to move on, for he was most comfortable when alone. Loneliness was unknown to him for he had been alone all his life; companionship had never been his to enjoy.

For the most part he survived off the land by hunting and trapping. What little he needed was obtained by trading off the hides and furs, either at the closest trading post or by a trek to a nearby settlement. If the year had been kindly and his labors brought more than needed for supplies he would indulge in a drink or two while enjoying a game of poker. Times were he profited by the hand he held, while on other occasions his pockets were emptied.

The winter past had been long and game scarce; although he did not have many pelts to show for his labor he needed supplies just the same. The settlement was closer, but Josef went instead to the trading post where there would be few people to encounter. The pelts earned him enough to purchase what was needed, with little left over. When the owner of the post suggested a hand of poker he hesitated, for he had no desire to lose his last dollar. But then he laughed at himself when he realized that he had no need for money; he already had his supplies, so why not enjoy a hand or two?

A shot of whiskey and a good cigar – he savored both as he examined his cards. Placing his wager, he drained his glass and sat back. He won the hand. His host poured them each another whiskey before dealing the next hand. Josef relaxed and settled into the game. After an hour or so another who had come to trade off his pelts joined them. With each passing hour, Josef's fortunes increased. The owner of the post had not been as lucky; his cash was depleted. He rose from his chair and poured another round of whiskey. He would wager his post against Josef's pile of money, he declared. Josef told the man he was drunk, and he had no use for a trading post. The game was over; he would finish his drink and leave. But the man persisted in challenging Josef to one last hand. Sighing, Josef could see no way to avoid it and agreed. When the hands were revealed, Josef became the owner of the trading post.

He did not enjoy the winning; it made him most uncomfortable as he watched the former owner prepare to leave the following morning. When Josef asked him where he was heading he replied that he thought this to be a fine opportunity to meet his grandchildren. Josef remarked he never knew the man to so much as have a wife. His wife died the winter before Josef first arrived, he revealed. Their only child, a daughter, had married several years previously. He and his wife had set aside savings when the trapping was still good and, as a wedding gift, they had given the couple money to purchase some land. His daughter had written to them over the years but they had never gone to see her. Her last letter had arrived a couple of years before, and he felt that this was as good a time as any to go and meet the grandchildren she told of in her letters. "And so where is it you are heading?" Josef asked. He wasn't sure exactly – her letters were postmarked some town in Utah. He figured he would ride up there and someone would be able to point him in the right direction.

Before the man rode off Josef pointed out that he knew nothing about being a trader, and asked for some guidance on what to do. He had laughed at Josef and told him he didn't need to know anything.

"When someone comes in with pelts, you make them an offer according to how much you can afford," he explained, "If they need the money, they will accept. If not, well, you can either forget it or try to barter, it's all up to you!"

Now Josef stood and looked around his trading post.

"What the hell am I doing here?" he thought. "Here I am, a man of property; ain't I the fine one!"

He thought about simply packing up his horse and leaving, riding back into the mountains where he was comfortable. But then he remembered the winter past; it had been a harsh one and days often passed without any game. He would give it a year, he decided. If nothing else, he could always lose it in a poker game, he chuckled as he started to survey his new home.

The summer passed quickly. Josef discovered how to barter; what the former owner had told him was accurate. The first time he made an offer for a pile of pelts, he expected the man to take them and leave. Instead he began to barter with Josef until they finally reached an agreement. Each time it became easier and Josef grew more skilled in assessing his customer. His greatest surprise was the arrival of a wagonload of supplies. He had never given thought to the day when his shelves wouls be emptied, and that they would need to be restocked. The previous owner had ordered the supplies months before, but before the unloading began he was told that payment was still owed. Certainly he would need the shipment, so without argument he paid the bill in full.

Josef helped with the unloading and, when done, offered the driver a shot of whiskey. As the two men relaxed, Josef related how he had won the trading post. Realizing he would be in need of further shipments, he inquired how to go about arranging for them. After the man left he wondered what other surprises awaited him, what else remained for him to learn.

The morning air now held a chill. He had gone out to fetch wood for the stove and he stood now looking up into the mountains. Already the aspens shimmered like gold in the sunlight and he longed to be back, riding along some stream in search of game.

"Why not?" he thought. "I am my own man, there is no one telling me I must remain here. It is for me to decide." Surely, he could be gone a day or two – or possibly longer.

It felt so good to be back again in the forests. He had wasted no time once he decided to go; he had simply packed his horse, shut the door to the trading post and rode off. Now he found more pleasure in the riding than any thought of hunting or trapping. He allowed the horse to pick the trail while he relaxed and thought of the many changes his life had experienced. He recalled standing with his father and looking at the snow-capped mountains in the distance. He wondered again what had been his father's fate. Had some accident befallen him, or had life become too hard for him and he simply rode off in search of something easier? No, he decided, he doubted his father would have abandoned his family. There is no ease in life, all one can do is face what is in front of them, however harsh it may be. There is no getting around it, just like these mountain trails, you move along without knowing what waits up ahead.

After two days he was ready to return to the trading post. He had lost his desire to kill an animal when he had no need for either the meat or the pelt. Realizing he was riding without reason, he turned the horse and headed home. Upon his return he discovered it had been visited during his absence; the visitor had left money to pay for what was missing from the shelves. He had not considered the possibility of someone needing supplies during his absence and he thanked his unknown caller for their honesty. More importantly, he became aware that his trading post was needed. People depended on the supplies available here. He was unsure how he felt about this awareness and decided he had time enough to ponder it once the winter snows set in.

The years at the trading post had made him soft, he realized. Too many years lived with the comforts of a roof overhead and a bed to lie in. He doubted his body would again welcome living as he once had; his movements had slowed and some mornings the stiffness of his body slowed him even more. How quickly those years had passed! The fur trappers were gone; it had been several years since he had last bartered over the worth of a pelt. He discovered he missed those friendly encounters with those fiercely independent men and he wondered where they had all gone off to.

Time had changed his world. More people had migrated to the state with each passing year; soon it would be so crowded there would be no space to breathe, he thought. Most of his customers now were faces he had never seen before and probably would never see again. The settlement had grown and most people preferred the stores there. Often, days would pass before his old dog would bark to announce someone approaching.

"Even the dog has grown soft," he realized. "He now prefers to spend his days lying in the sun and his nights on the rug by the stove. Perhaps there is some ease in life; it comes when we grow too old to care what lies beyond the next bend in the trail."

Summary

The beauty of life is that nothing is ever wasted. Everything you do serves a purpose. Everyone's life is an interesting story, especially yours. It's impossible to recognise that fact if you don't examine your life. Your unexamined misconceptions and expectations create the illusion that the lives of others are more interesting than your own. So, you pay more attention to others and wonder why life is so difficult and unrewarding.

One way to appreciate and understand your life experience better is to write your own biography. The process of writing it allows you to see things about yourself that you were previously unaware of. It provides you with numerous opportunities to examine things that you didn't want to previously look at for one reason or another. Your completed biography can become a family heirloom that others in your family will enjoy.

There was much about Angel Mara's life experience as William that remained secret throughout his lifetime. Secrets are heavy burdens for those who carry them; they can also serve to imprison everyone concerned.

Sir Thomas Bodley was William's father. He and William's mother had enjoyed a loving relationship but when William's mother became pregnant, he was unable to marry her. However, he arranged for William's mother to marry so that William wouldn't be born a bastard. William's mother's cousin had both title and estate, but he was without finances and marriage to a cousin, already with child, was made more attractive by the funds that were made available to him. This man is not to be faulted. He loved William as much as he loved his natural born son, James. However, James was a constant reminder of the intellect of a son born to him. Therefore, he was continuously reminded that William wasn't of his blood, no matter how much he desired it differently.

It was Bodley who visited William when he was in London with his parents. Although Bodley never acknowledged William as his son, he did retain control over certain aspects of his life. It was Bodley who determined that William would attend both Eton and Oxford. Thus it was also Bodley who paid for William's education.

He loved William as much as a man can that must witness his son being raised as another's; his pride was evident as he appeared both at Eton and Oxford to witness William's growth. William's entrance into parliament was marked with the gift of a very costly watch. It is a pity that circumstances prevented him from experiencing the joys of true parenthood.

William was not raised without love, even though the truth of his birth was kept from him. In turn, William loved his stepbrother, James, very deeply and it was most difficult for him to accept that James' mind would never grow into adulthood. The love both William and Mary Elizabeth gave him was returned over and over as William continued to care for him until his death. William's young children mourned the loss of their uncle; he was their favorite playmate.

What purpose did this all serve? Certainly one must recognize the presence of tolerance and acceptance in the relationship with James. It is easy to love a child; it is not always easy to continue loving a child in the body of an adult.

This incarnation was also an examination of the Law of Cause and Effect or what is often referred to as karma. The keeping of the secret created an effect (karma), and those keeping the secret must now deal with it.

It is impossible to say if William's life would have been different if he had known the truth about his father. What is known is that William enjoyed a life that would have been denied him had he been known as a bastard son. That was not William's responsibility; he played no part in his conception or in the secret. There was no guilt for William to assume for the life provided him by the secret.

Unfortunately, this personality is unable to accept that, just as it was unable to accept James' condition. James was the product of a bloodline weakened by intermarrying. Unable to care for himself, he was surrounded by those who loved him up to the day he died. The nurturing he received was of greater value than title and property combined. Blinded by guilt, William was unable to understand this.

When this personality became aware of the secret that had been withheld, it felt it had inherited title and properties that should have gone to James. It falsely accepted responsibility for robbing James of his birthright, and the guilt it assumed at that time was overwhelming. Consequently, life veiled this incarnation.

Most Americans admire their founding forefathers. They admire them because biographers and historians gave their lives purpose, meaning, and value. Yet none of the founding forefathers saw themselves as being different, special, or someone to admire. They were merely people, living their daily lives with the same concerns we each still have today: the safety of their homes and loved ones, as well as the taxes they were paying. Not too much has changed, has it?

They were too involved in their daily living to see the purpose, meaning and value of their lives. This fact is clearly illustrated by Angel Mara's life experience as John Quincy Adams. This incarnation was most assuredly an examination of responsibility. It is a thread that is woven throughout this narration and weighed heavily upon him from a very young age. John felt responsible for his mother and sister in his father's absence. As a child he became responsible for the bird, and as an adult he assumed the responsibilities of governmental service. He had been so busy being responsible that he hadn't taken the time to sort out whether the responsibility belonged to him or to someone else.

The life experience of John Quincy Adams clearly demonstrated his intelligence and ability to adapt. Angel Mara demonstrated these same abilities and qualities again as Josef, the mountain man.

Do you wonder what could possibly be gained by leading the isolated life Josef did? Why would the personality choose to experience an incarnation so lacking in personal relationships? Keeping in mind that "everything serves a purpose," we must examine this experience to discover its hidden value.

Josef was never lonely, because he had never experienced companionship. You cannot know or miss what you have never had. If you've never known anything other than loving relationships, how do you come to fully appreciate and value their worth? If you can understand that concept, you can see that the personality chose to experience the lack of relationships so that it may come to a more complete appreciation of how they assist in one's personal growth.

The personality was actually examining life from an opposing position: whereas in the previous incarnations, there were loving family relationships, in this one there were none, not even any siblings.

Previously, education was experienced, but Josef remained without formal education. He remained unmarried and childless. But he was not unhappy. I sensed contentment about him – the life he lived fulfilled his needs and that was all he asked or expected from life.

This was also an examination of self-reliance. Again, in the previous incarnation presented, this personality was born into a family of means and position. His education was provided for him; he never really knew what it meant to go without. True, he made his own way in the world, but he didn't have to be totally self-reliant like Josef.

A Veiled Incarnation

The source of this personality's problems and conflict is its inability to accurately discern and unconditionally accept what life presents it. The ability to quickly accept and wisely respond to what you are experiencing is responsibility.

William's strong will and determination prevented him from completely accepting his brother James' condition and destiny. William wanted things to be as they were when they were children. As an adult he wanted James to be his equal, which was impossible.

In the next incarnation life assisted this personality (John Quincy Adams) in seeing that James didn't harbor any resentment towards him. In that incarnation James' personality incarnated as Abigail Adams, John's mother. Examining the relationship between John and his mother allows you to see that James' personality benefited from the love and devotion given to it by John in the previous incarnation. The actions of Abigail Adams clearly reveal that this personality harbored no resentment or ill feelings towards William, and that William's perception concerning robbing James of his title was incorrect. Had it been true, Abigail would have responded to her son John differently. Life assisted this personality in seeing that again today. It can easily see this by observing her relationship with her present lover, friend, and partner.

The perceived betrayal by William's father was traumatic and caused this personality great pain. Upon passing into the non-physical realm the personality was unable to relive the pain and trauma so that it could review and reflect upon the experience. Had the personality done so, all memory of the experience would be gone now. Furthermore the personality would be benefiting from the understanding it gained.

The personality has difficulty reconnecting with an incarnation that has been veiled because the veil serves to disassociate the personality from the experience. The personality must associate itself again with the experience for a complete healing to occur.

In today's society it is easy to become so overwhelmed by what you're experiencing that you're unaware of your emotions. When that happens it is a good time to step back and ask yourself, "What am I feeling now?" Keeping a journal or diary of the various emotions you experience will assist you in seeing which emotions make you feel comfortable, excited, and motivate you. Likewise, note which ones cause you to feel uncomfortable, depressed, and rob you of your energy. The same applies to expressing your emotions. Which ones are more difficult to express than others?

Chapter Seven: Angel Anne

Carausius

It was late in the third century, and the Roman Empire was jointly ruled by Maximian and Diodetian. Maximian's campaign against the Bagaudae rebels in Gaul in 286 had been successful. His co-emperor, Diocletian, established an autocratic government and was responsible for laying the groundwork for the second phase of the Roman Empire, which is known as the "Dominate".

Life was difficult during this era; the lives of those in power were often cut short. Fear and the desire to preserve their longevity caused most citizens to avoid life and remain in obscurity. Few possessed the courage and intelligence to step forward and face the perils of life.

Somewhere outside Rome, a lone rider emerges slowly from the mists of the heavy fog upon the land. Both horse and rider appear exhausted as the horse cautiously finds his way on the dampened roadway. Stopping finally, the horse paws the ground, snorting, and the rider is awakened from his slumber. Quickly glancing first one way and then the other, he realizes why this poor dumb animal so rudely roused him. Here, the road is no more!

Tired and without patience, he dismounts to relieve himself. Hunger is rumbling in his stomach and sorry he is now that his departure from Rome was so hasty. "Damn them all," he mutters to himself, "If they had but a brain the size of a bird they would perhaps be able, just once, to entertain an idea not fed to them as food for a child. The fools agreed to rule jointly–they couldn't see that a throne divided is no throne. If there is no throne, there is no power!"

Where had he erred? he wondered. His planning had been so thorough: with the legions he, Carausius, commanded it would have been so simple to place one man upon the throne, one man who would then owe his crown to the one had who secured the title for him. Instead, the little boys grew afraid that they would lose their toys so they went running for help. He could see it now; they had no spine!

"Serves them right," he grunted. "No spine, no throne, no power."

It made no difference now; he was fortunate to have escaped with his spine still in one piece. They both would have seen him sent into the next world if they had caught up with him. Fleeing, he had not stopped to pack any supplies, but he had paused long enough to retrieve the bag of coins he had hidden. Their intended use had been to purchase favors, open a door when needed. They would not be needed for that now – a better use now was to protect his hide.

He was fully awake now. Knowing he must be on his way, he stood astride, hands on hips, looking at his surroundings. His weary horse moved a couple of paces, gave a low snort and nudged him.

He recognized the area. The landholder here was known to him, but he dared not approach for he no longer knew who could be trusted. "Only myself! Yes, you, you old warrior, and myself," he muttered to the horse as he scratched the animal's neck in contemplation. Food and rest first, then away from this land of spineless children, he determined.

Mounting, he turned the horse east. The sun would soon be rising to warm the land as well as his bones. To the east was a small forest where one could be assured of rousing some manner of game. There was wood aplenty for a fire and a stream for drinking. If memory served him well, there was a ravine also which would conceal man and horse for a day of needed rest.

Thus the career of Carausius in Rome was at an end. He had plotted to set one man on the throne – it mattered not which one – but the two candidates had foiled his plot by agreeing to rule jointly. He was a bold man, unafraid of taking control of a situation once he could see what was needed. Quick to learn, he knew now where he had erred. He had sought power through the throne of another; he had not been bold *enough*!

Passage was secured on a ship bound for Britain and he spent the journey pacing the deck, contemplating his future. The coins, well concealed upon his person, he knew amounted to a small fortune and they would open the doors to his future in Britain. No more would he seek to open doors for another: he would seize the day for himself.

And so Carausius went about the business of establishing himself upon his arrival in London. It was never difficult to locate soldiers in need of a thirst-quenching in exchange for some conversation. Quietly, he went about his fact-gathering. He knew how to talk to these men. They were no different from the Legions he had commanded in Rome. Patiently, he began to form a plan.

He had taken quarters befitting a man of station. No credentials required. He had the coin; that was enough. His home quickly became a center for parties enlivened by debates that challenged the revelers to think. He drew to him a circle of supporters who enjoyed his strong presence. He made them think and feel that they could be greater than who they were and for this they were ready to lay down their lives for him.

Next, he courted the other wielders of power, those who possessed the wealth. One common concern shared by persons of wealth, no matter where they be, he knew, was that they had wealth and they wanted more. They had no desire to part with one coin unless it would gain them two. As he moved through their midst, a word was whispered here, another there, and he established himself as one of them. Marriage to a daughter from one of the finest houses in London completed the creation of his persona. Her father, a leading merchant, eager to expand, was ever searching for new trade routes. The marriage, beneficial for both men, was a good one. The bride, comely enough, was pliable and eager to please her husband, and that suited him well.

Painstakingly he had woven a plot, thread by thread, and his tapestry now was complete. Surrounded by those who controlled the wealth of London as well as supported by most of the troops garrisoned there, he was ready to make his move. And a bold one it was!

He simply strode into court and proclaimed himself Emperor of Britain! Sword in hand, he stood awaiting any who dared answer his challenge. But he knew what would happen. Yes, once again those seated by Rome were without spines! They dared not rise to the challenge; instead, like mice they quickly scurried off lest they be seen and captured. "Save your worthless hides," he shouted after them, "It matters not for I have seized the day and Britain is mine!"

It is only fair to say that Britain prospered by the hand of Carausius. The treasury grew fatter and was further enhanced when he established a mint. Allectus, his minister of finance, was greatly pleased by this expansion of his position. Perhaps too pleased, Carausius pondered. He decided to keep a vigilant watch to determine if it would be wiser to place the mint under the control of another – create a new minister for this fledgling department. Power divided is no power! Yes, he must remember that. The treasury was enough for Allectus; the mint will have to go to another, he concluded.

His head hurt from all the problems of the day. As he stood, shaking as though to free himself from the day's affairs, he was suddenly taken by the memory of his journey to this moment and he smiled to himself. "I was right, old warrior," he found himself saying. "Trust only in self and you, my old steed," remembering the oath he had muttered to himself what now seemed so long ago.

Wearily he made his way up to his bedchamber. He had brushed aside his aide's offer to accompany him. He knew his wife would already be retired in her chambers, but decided to pause and speak with her a few moments. This was something he was learning to take pleasure from. She was now thick with child and he found her company to be calming. She would chatter on about her day, filling him in on court intrigues that women always seemed more aware of than men, and this was a respite from the problems he knew would await him in the morning.

He rapped upon her door, as always, awaiting her invitation to enter. Hearing no response, he rapped again. The hairs on the nape of his neck bristled as a cold chill swept over his body. He paused but one brief moment before bursting through the door. The scene repulsed him; crying out, he attempted to wipe its horror from his eyes. Blood everywhere! His wife lay dead, disemboweled as the child was cut from her womb. The infant was held up before his eyes before they sliced it in half. The next swing of the sword severed his own head from his body.

It was a kindness that Carausius was killed. Although he was bold in his actions, he was not cruel in nature. Thus, the horrific manner used to slay his wife and child were more than he would have been able to live with, and the scene would have haunted him forever and eventually driven him mad.

His assessment of Allectus had been accurate; the observation, unfortunately, came too late. Allectus, now enjoying the expansion of his powers, knew Carausius all too well and knew it was a matter of time before he would find a way to limit his power. He merely decided to act first, and hired mercenaries to dispose of his opponent.

Catherine, Sister of the Abbey

We return to England again during the fourteenth century. Young Richard, the second son of Edward the Black Prince, is the King of England. Richard's brother, the Duke of Gloucester (Thomas of Woodstock) and others are plotting to overthrow his brother. Thomas of Woodstock and other ringleaders of the Peasant's Revolt of 1381 are arrested and executed. Richard's disingenuous tactics disperse the rebel forces from the streets of London and end the disorder.

In the examination of this incarnation, it is at first difficult to ascertain purpose. At times the role the personality is attracted to is so well disguised that it obscures the experience that is being sought. As during the previous life experience, this is an examination of faith, character and power. This wise personality was able to reconcile the conflict between her religious instructions and those of her psychic visions. She accepted her position with grace and dignity, even to the end of her life.

Now, there in the rolling meadowland are several large stone buildings, the largest of which is a church. You can hear the nuns as they sing their evening prayers. Also present are their students: young women entrusted to the good sisters to be educated and kept in purity until they are of an age whence their fathers or guardians can arrange a favorable marriage.

And there is another present. Sitting in the last pew is a solitary figure clad entirely in black. Head bowed, the folds of her hood hide her face and she appears to be praying as she fingers the beads in her hand. As we draw closer it becomes apparent that she is not following the evening service, but is instead lost in reverie.

A smile plays now upon her lips and softly she can be heard as she says, "They call me the old crow. If only they knew the path this old crow has flown. More have these old eyes seen than they will ever know."

She sees herself once again as the tiny young girl her father had brought here when she was but seven years of age. She remembers again how frightened she had been. Her mother had died, taken by the Black Plague. This much she knew because the cook all too willingly had told her it was because her mother had not laid aside her wicked ways and asked for forgiveness. Instead, she remembered the cook telling her that her mother had continued the old ways, boiling weeds (herbs), throwing open the shutters, letting all that pestilence in. Cook had never liked the girl's mother and she filled the child with all manner of tales now as she readied the child for travel. Her father had left instructions that upon his return, she be ready to leave London.

Sitting there, she recalled the fast-paced journey. Her father had said nothing to her; he had simply motioned her into the empty carriage. It was dusk when they arrived here; she heard the voices in the distance singing the evening prayers before she could see their journey's end. She sat in the carriage, not knowing where she was or what awaited her. She had dozed off and was finally awakened and taken to the kitchen for some bread and warm milk before being taken to the maiden's chambers.

Here, then, she spent the next six years, learning the prayers and observances so necessary to one who desired to save their personality from the ways of the devil. As hard as she tried to be good and to remember always the teaching of the priests, she would find herself asking some question that angered them. Or she would share her dreams with the Sisters, only to find herself doing penance. Early on she had discovered it was best to keep her visions apart from her lessons; only once had she tried to describe one of them. She learned quickly that visions were the devil's work, as an attempt was made to starve the demon. Hunger is a fast teacher, she discovered.

As she approached her fourteenth summer, her father sent a carriage for her. As suddenly as she had taken up residence here, so too did she depart. She found herself returned to London, and she remembered it not. The house of her father was many roomed and she recalled wandering from room to room, looking for some familiar thing from her childhood. But there was nothing! All that had been her mother's was gone, removed as though she had never existed. As she moved about the house, her visions allowed her to see what she knew was not there. First was the light, so bright it would blind her. Then from the light would come the vision. Often, it would bring her mother to her. She knew the priests would have instructed her to fall to her knees and recite her prayers to drive away the visions, but she could not see any evil in them. They did not frighten her as the threats of eternal damnation frightened her. Instead, the visions left her feeling warm and cared for.

Her father, gone from the household most of the time, had secured a female companion for her and the two quickly became friends. The companion, Margaret, assisted her in assembling a wardrobe of gowns to replace the simple garments she had worn while at the Abbey. Margaret then proceeded to school her in the social graces, as well as the art of flirtation.

It was from Margaret that she learned the truth of her heritage. Engaged in the chatter of females one afternoon, Margaret had unwittingly referred to her mother as a mistress. When pressed for an explanation, Margaret explained to her that her father was Thomas of Woodstock, the earl of Gloucester, and had another household elsewhere. This house was the house he had purchased for her mother, whom he had never wed. She had never thought about who her father was; he had always been simply her father. She had never considered his life beyond the role of her parent. Now she recalled how Cook had called her mother wicked and evil, and she understood the full implications of those words. She was a child born outside the sanctions of marriage. She had no claim to her father's name nor his properties. In actuality, she was at the mercy of her father's conscience and generosity, a position she found to be most uncomfortable.

Her father arrived late one afternoon and she found him deep in thought in front of the fire in the great room. He had requested her presence, and she patiently dressed her hair and selected a gown that was most becoming. As she entered the room, her father glanced up, startled, started to rise and then re-seated himself as though unable to stand.

The two of them talked into the night. He disclosed that upon her entrance, he thought it was her mother coming to him again. He called her mother Kate and he told her that she had received her own name, Catherine, from her. He had been unaware of the strong resemblance she bore to her mother until now, when she had reached maturity. With this knowledge, she questioned him about the manner of her birth as well as her future. Her father assured her that, although he was unable to publicly acknowledge her, she would always be well provided for. This house was now hers as it had been her mother's before her, along with an annual stipend for her personal expenses. He would continue to provide for the household expenses as well as the staff he had secured to provide for her wellbeing.

Never had she enjoyed the intimate company of her father so, and she became aware of the deep affection he had for her mother as well as his concern for herself. It really did not come as any great surprise when he finally spoke of the visions that came to her mother, and she was fascinated to learn that often the visions had provided him with direction and purpose. She now understood how completely she was her mother's daughter, and she confided her visions to her father. For the first time she was able to speak of them without fear of reprimand, and it was to her a great relief to share what she had been taught to hold secret.

Her life no longer bore any resemblance to her life with the nuns. Margaret introduced her to a social life along with the intrigues of court. Several young gentlemen sought her company and she discovered she enjoyed their attentions. Knowing she was not acceptable marriage material, she did not take their attentions seriously. Thus, she appeared aloof and unattainable, which only served to heighten their desires. With Margaret as her tutor, she learned to play the social games with flair and expertise.

Often her father would arrive, late at night, seeking her to call upon her visions. Through these late night discussions, she learned of her father's involvement with others as they attempted to express their opposition to the King's (Richard II) evolving tyrannical measures. She shivered as a cold foreboding cloak seemed to settle upon her and she warned her father of the danger of his involvement. No longer were her visions to be enjoyed, for they now foretold of the extent of the King's tyranny as well as the measures he would take to preserve his throne.

But her father told her he must continue with what he had started because the clock cannot be turned back. More often became his visits, and her house was busy with the activities of his comings and goings. He shared not the details of his activities with her and she often found him in the company of men unknown to her. When she spoke to Margaret of all this, Margaret attempted to allay her fears and suggested they journey to the country for a visit with Margaret's aunt.

It was here then that her father's courier found her. Instead of a distracting country weekend, she now was confronted with the collapse of the life she had come to accept. The packet the courier gave to her contained her father's letter as well as other documents. The king had denounced him, along with those who stood with him. It was uncertain if he would be banished or arrested and imprisoned. But in either case, his title and properties would be surrendered to the crown. Tales were being told of her involvement and he feared also for her life. He encouraged her to seek sanctuary among the Sisters who had raised her. Begging her forgiveness for involving her in his opposition, he instructed her to deliver the documents he was enclosing to the Mother Superior upon her return to the Abbey. These, he told her, would release funds he had secreted away many years ago, and would allow her a life of dignity among the Sisters of the Abbey.

Thus, she once more traveled the countryside, arriving in the meadow at dusk. She heard the familiar voices lifted in their song of prayer. It welcomed her, comforted her, and she knew she had returned home.

So, now she sat here listening to the voices lifted again in that familiar prayer. Long now had she been able to reconcile her visions with her understanding of God and they no longer came to her – she knew that they were no longer needed. They had served as her protection and she knew also that they had been sent by a mother who loved her daughter even beyond this mortal world. With that knowledge she found a contentment and peace unknown to her before, and she lived out the final years of her life at the Abbey in quiet contemplation.

Summary

Occasionally I run into something that really puzzles me and leaves me speechless or close to it. I hate to admit it, but the latter is rare and it is something that I am working on.

Something about Angel Anne initially stumped me, and I was unable to figure out what it is was or express it. My impression of this personality is that it is a very unusual angel, in a very favorable way.

At first it appeared as if Angel Anne had been examining power from numerous perspectives. That puzzled me because I never encountered an angel examining power as an objective: angels usually do that long before ever becoming an angel. It isn't until you look at the life experiences of this angel differently that you see that it has been examining adaptability for a very long time and has become a master at it. Adaptability also involves responsibility because it is the ability to quickly and accurately discern what one is dealing with and wisely respond. You cannot do that unless you are very flexible and able to quickly adapt.

At first it's difficult to ascertain the purpose of Angel Anne's incarnation as Catherine, daughter of Thomas of Woodstock. That's because the personality's role that it's attracted to is so obvious and it is the obvious that disguises its purpose, objective and value.

Although Catherine was confronted with the conflict of her religious instructions as a nun and her psychic visions, she managed to reconcile the two. She denied neither and found value in both.

You have to admire her ability to quickly adapt to her role as a nun at the Sisters of the Abbey, an entirely different environment than the one she had been used to. Later, as a young woman, she adapted to the role of being the daughter of Thomas of Woodstock, the Earl of Gloucester. She accepted that she was born out of wedlock and had no claim to her father's name or his properties, even though she found it uncomfortable having to rely upon the generosity of her father. Soon after that, she became her father's advisor. All this changed later on and once again she found herself back at the Abbey. She accepted her position with grace and dignity, even to the end of her life.

If you examine this personality's earlier incarnation as Carausius, you will see that he was also very adaptable; he couldn't have accomplished what he did otherwise. The internal strength of this personality is displayed in both incarnations.

Angel Anne's life experience as Carausius was veiled because of the horror Carausius saw, and his unexpected murder. If you recall, Carausius rapped upon his wife's door and awaited her invitation to enter. Hearing no response, he rapped again. The hairs on the nape of his neck bristled as a cold chill swept over his body. He paused but one brief moment before bursting through the door. The scene repulsed him: crying out, he attempted to wipe its memory from his eyes. Blood everywhere! His wife lay dead, disemboweled as the child was cut from her womb. The infant was held up before his eyes before they sliced it in half. The next swing of the sword severed his head from his body.

It was a kindness that Carausius was killed. Although he was bold in his actions, he was not cruel by nature. Thus, the horrific manner used to slay his wife and child was more than he would have been able to live with – the scene would have haunted him forever and would have driven him mad.

I am pleased to say that this personality is successfully dealing with the trauma experienced as Carausius. This angel is definitely a role model in adaptability.

I didn't mention this in the stories about Angel Anne, but a later incarnation of this personality allows you to see how the personality seeks balance in what it experiences. He was known as Thomas Townshend, a former Chancellor of the Exchequer – the British cabinet minister responsible for all financial matters. He was known for his pointed wit and polished eloquence. Although one of the most accomplished senators that ever sat in Parliament, he would change his political position almost monthly. Many of his speeches were unrivalled in parliamentary history for wit and recklessness. He was responsible for passing the Townshend Act – a tax on glass, paper, and tea levied on the American colonies. Some of his colleagues accurately predicted that his act would cause England to lose the American colonies.

The heart of this personality is in the British Isles and that is where it keeps incarnating. Angel Anne is presently deeply involved in a review incarnation and is experiencing some difficulty adjusting to not being involved in public service.

Humans are creatures of habit; familiarity makes them feel comfortable and secure. Before they do something they want to know what they will experience. When was the last time you did something spontaneously? How did it make you feel? Now compare those feelings to the ones you experience when you do something familiar or repetitive. Which feelings did you like better? Maybe it's time to spice up your life a little by being more adventurous and spontaneous!

Chapter Eight: Angel Nan

Attorney Jan of Belgium

This life experience took place during the thirteenth century in Belgium. It was a time when attorneys were not the respected professionals that they are today.

On several occasions life presented this personality with obstacles that caused him to pause and question the purpose of his life. His ability to examine and discern were two of his three greatest assets. The success he achieved was due to his ability to sift through what was presented so that he could determine the actual issue.

To always respect the dignity of another is an ideal few concern themselves with, but for Jan, it was a way of life. He was unable to see his own value because it was as much a part of him as breathing.

Herein we also find the purpose of this experience for the personality was examining the purpose for its own existence.

That Jan's father send him away to complete his education had appeared wise at the time: too often tutors were no more than men with little knowledge, unable to make their way in life. Already he knew more than his latest tutor – it would be better for him to go where learned men could be found to teach him what he did not know.

Now he wondered what purpose his education would serve. Six years he had been gone, returning home for holidays when travel permitted. It had been a year since last he traveled these roads, but as the countryside grew ever more familiar, his heart grew heavy. His homecoming should have been a time to celebrate, but instead he was returning to mourn. His father had passed on; as the only remaining heir it would fall to him to see to the burial and settle the accounts.

Taxes; damnable taxes! They left him almost nothing. The small estate, which had been in the family for several generations, was gone: gone to settle his father's accounts and the taxes. No fortune remained for him to watch over, to increase through land purchases and wise management. The family name would vanish: no longer would any land be called Van den Holst.

Jan removed little from the home: the portrait of his grandfather that had hung over the mantle; silver candlesticks; his father's gilded money box, and a few other trinkets. Family friends in the nearby village offered him a room in their home, where he was welcome to remain as long as needed. Without their kindness his future would have been bleak – the few coins remaining would not have sustained him for long.

The following year presented little opportunity, no matter that he was an educated man. He grew ever more despondent and withdrawn until his well-being became a grave concern for those he lived with. The priest was asked to call, to counsel him and pray with him for God's guidance.

Gently Father Xavier spoke with Jan until he could bear no more. Angrily he lashed out – not at the priest, but rather at his circumstances. It was all so futile. His efforts earned him nothing! Although educated, he survived on the kindnesses of others. What purpose could God have for him that he must endure such an existence?

"Perhaps it is meant for you to become a tutor, Jan, to take what you have and share it with others," the priest suggested.

Remembering the tutors of his boyhood, Jan shuddered. In memory they seemed not as men but rather as shadows shuffling through his life – shadows leave no footprints, no lasting impression to evidence they once had passed that way. Was this to be his future? If so, it was a future he found without promise.

"What God intends for you is known only to Him. As mortals, we can only pray that He will guide us that we may come to know His intent. I doubt it is intended for your education to be wasted. It is for you to discover how you may share what you have been given."

For many days Jan pondered on the words of the priest, and his visits to the church grew more frequent. Kneeling there in prayer, he began to know peace once more. It surprised him to discover his heart now dared again to contain hope, and from that hope came an idea.

Visiting the merchants in the village, he made it known that he was available to advise them. Should they have any documents that required reading, he could do so and explain the contents to them. Should they need to send any communications, he could draft the messages and see to their transmittal. It was a tedious task, as most of the merchants were uneducated men and had difficulty understanding that these were services worthy of financial reward. Patiently, he explained the importance of understanding any agreement they may enter into and the costs a misunderstanding may bring upon them. He developed an ability to explain these issues in a manner that was easily understood.

Jan became a trusted confidant to many and learned more than needed while settling their disputes. Upon learning the nature of their problems, Jan would refer some of his customers to the priest for guidance. Jan listened very attentively to everyone who sought his guidance, remembering how Father Xavier had responded to him in his time of despondency. Trouble can rob a man of his dignity, he realized, for it humbles any man to ask another for help.

A father and his lame son appeared at Jan's door, sent by the priest, they said. They carried with them an old document granting the family rights to the coal that lay on land owned by the Crown. All who worked these fields knew about those rights, as had their ancestors before them. Selling the coal was their income, but now men had come and driven them off the land, and threatened them with clubs so that they were afraid to return. How can others claim the coal that is rightfully theirs, they asked?

"Many of the old grants are no longer honored because those that had awarded them are no longer living," he forewarned the men. The only way to determine if the rights awarded in the document were still theirs would be to present their dispute in Brussels. Jan advised them to travel there and employ a man such as himself to argue their dispute.

To search out someone new would take much time, and most were unwilling to listen. They asked Jan to represent them, inquiring what his fees would be. Arriving at a sum agreeable to all, he and the father made the journey together, leaving the son with Father Xavier.

His return to Brussels awakened old memories, for it was here he had received his education. Those had been carefree times, he recalled as he saw again streets familiar to him. When unable to travel home for a holiday, he and others who remained lodged in the city had wandered these same streets in search of adventure. Recalling their escapades brought a smile to his lips and he wondered what manner of men his old friends now were.

Knowing his client had limited funds, Jan located inexpensive lodging for them and began his struggle with the bureaucracy. Often he found himself feeling as though he was wandering a maze, referred first to one office only to be sent to another.

When at last the opportunity to present the dispute arrived, his client accompanied him, carefully guarding his precious document. Jan presented his client's case with passion; the grant had been awarded as an honor. It rewarded this man's family for their loyalty and service. Already their own lands had been taken from them; now this was threatened as well.

It was several days before they discovered the reason for the dispute. The land had been sold and the new owner had the right to be present to argue his rights to the coal. Notification had been sent and they now must await his arrival.

During the wait, Jan revisited his youth by wandering the familiar streets. He even enjoyed a reunion with two of his former classmates, spending a boisterous evening reliving many of their adventures. Both men were engaged in business ventures here in Brussels. Their manner of dress and carefree spending of money evidenced the success they were enjoying.

The new owner of the land arrived to present his claim to the coal. To this man and his agent it was clear and simple: When a man buys land, he also buys whatever exists upon that land. If it were a house, the house is then his. If it were trees, the trees become his. It is never said that you may buy the land, but you may not plant the soil. The coal was his; these men were stealing from him.

"Why was any examination of old documents neglected until now?" Jan argued. "Why was the land sold without notification to my client?" Had the records been examined, the existence of his client's grant would have been discovered, requiring that he be notified and given the opportunity to purchase the land himself. His client should not suffer because of improprieties committed by others.

Therefore, the existence of the grant demanded his client retain the rights therein.

Again a wait was required; time was needed to investigate the dispute. Their expenses continued, however. Observing his client's demeanor, Jan cautiously inquired the reason for his despondency. After much encouragement, the man confided his funds were depleted. He could no longer remain in Brussels, yet he could not rightfully return to the coalfields. He would have to abandon his claim and return home.

Jan pondered the situation for a long time. This was a man with nothing but his honor and his good name. Any assistance he offered must allow his client to retain his dignity, he realized.

After some time, Jan presented a solution. "A man and his client may arrange whatever financial arrangements they both find agreeable," he explained to his client. "Times have been when I have funded the examination of the dispute, receiving repayment later when the client's solvency was restored. It is often more advisable for me to assume payment of all fees and costs for it perhaps enables me to negotiate and reduce some charges. If it were agreeable to you, we could enter into such an arrangement. When this is resolved and you are returned to your coal fields, you can repay what is owing."

That this was difficult for the man to agree to was apparent. He queried Jan at length. What if his grant were no longer honored? He would be left with neither the means to provide for his family nor to repay the debt. Jan could offer no reassurance, and he also wondered what the future held for this family should they lose their claim to the coal. The discussion continued late into the evening, until at last, offering his hand in agreement, his client vowed he would not rest until he was again free of all obligations.

The decision favored Jan's client. Although his opponent had purchased the land, the rights to the coal remained with his client as long as any heir remained to work the coalfields.

Jan remained in Brussels, deciding to allow himself a few days' leisure. His client had departed immediately upon learning he could return to the fields. Jan realized the debt weighed heavily upon the man and he knew that his client was eager to begin the repayment, and they parted in friendship. Now he visited again with old friends, enjoying the praise they lavished upon him for the arguments he presented before receiving the favorable decision. A man with his abilities could go far in Brussels, they informed him. Many already knew his name. He would encounter little difficulty finding clients, for the disputes were many.

It took little to convince him. He returned to Liege merely to gather his belongings and settle his affairs there. Before departing, he stopped for one last visit with Father Xavier; he could not leave without saying farewell, for he had grown most fond of this priest. He was sent off with assurances that prayers would be offered for his success and happiness.

The move to Brussels introduced Jan to a life filled with activity, and his reputation for enthusiasm and honesty earned him a large clientele.

As promised, payments were received with regularity from the client who had won him his reputation. As his prestige and fortune grew, so did his social invitations. It was the daughter of a wealthy client who finally captured his eye, and he pursued her with passion. He was not her only suitor, however, and he puzzled over how he could win the lady's heart. His competitors lavished her with costly gifts and he searched for something even more lavish and beautiful.

Everything he found seemed paled by her beauty, inadequate of expressing his love for her.

Remembering the guidance he had received from Father Xavier and through prayer, he turned to God, "If it be your will, Father, allow me to see what I have that no other can offer."

He pondered on his prayer for the words had come to him of their own accord.

"What do I have that no other can offer?" he wondered. In his thoughts he searched through his many possessions, discarding each idea one by one. What he owned could be possessed by any other man; he had nothing to give that could impart his love to her.

His old despondency returned. Father Xavier was not here to counsel and comfort him, nor did he find peace in the time he spent with God. Again he questioned the purpose of a man's life. If our heavenly Father is a loving Father, how then can he intend for man to suffer so? Has he allowed me to enjoy but a brief moment of success only to return me to such suffering? In anguish, he realized that his conversations with God had grown infrequent as his successes grew, and now he returned to God because he was suffering.

"Is this the way of man?" he wondered. "Are our sufferings necessary to remind us that we have given greater import to our affairs than we have to God?"

Too many questions that went unanswered. However, he had an engagement with the young woman. It would be rude if he did not keep his appointment. Bearing only a single rose, he called upon her.

"It had been my intent to search the land over for a gift which could express the depths of my devotion to you. I have found nothing, for there is nothing made by man that equals what I feel in my heart. I bring only this rose for in its perfection we can see the work of God."

They were wed. Their children loved the story their mother told of how their father had won her heart. Any man can bring jewels; only a man in love humbles himself as their father had, she would tell them. Their sons she encouraged to follow the example of their father when it came to affairs of the heart; their daughters she taught to look to a man's heart rather than his possessions.

Jan was saddened to learn of the passing of Father Xavier. He had not returned to Liege since his move to Brussels, something he now regretted. He must return to Liege – he needed to kneel in prayer in the church that had brought him peace so many years ago. In memory of Father Xavier he intended to gift the church with a large sum; perhaps they could establish a memorial befitting the life of that gentle man.

The passage of time was unapparent in the village; a few new homes and proprietorships in new hands. However there was little evidence that the prosperous growth of Brussels was also enjoyed here. He became aware that he felt as a visitor here – no longer did this village arouse the feeling of homecoming. Best he attend to the reason for his visit that he may return to the city he now called home.

The church doors stood open as they always had. Entering, he genuflected, acknowledging God's presence. Alone in the church, he knelt. He thanked God for sending Father Xavier to this village, for the guidance many had received from the priest.

"He has earned his eternal rest with you."

Ending his prayer, Jan rose. It remained yet for him to present his gift so that he may make ready for his departure. Wondering where he may locate the new priest, he stood glancing around. A movement caught his eye; turning, he saw someone approaching. Something about the gait of the man stirred a memory and he attempted to determine from whence he knew him.

As the priest neared, Jan asked, "Have we met before? You appear familiar to me."

He was the son of a former client, the man explained. He and his father had traveled to Brussels to present his father's claim.

All the old memories washed over Jan as he related how honorably the priest's father had fulfilled his obligation. With sadness, the priest related the events following his father's return and his death later.

"He was on his way here. Resting under a tree, he passed quietly."

How tragic! His debts were paid, he perhaps could have found a few moments of pleasure in his later years, Jan thought. Or perhaps the man found pleasure in fulfilling his obligation, in knowing his only remaining son served God. "Who am I to question whether a passing is tragic or timely, for it is all according to the intent of God," he realized.

Presenting the purse of money to the priest, he became aware that it lacked in comparison to what he had been given by Father Xavier. No amount of money could ever equal the value of that man's life.

The new priest accepted his donation. It had always been the desire of Father Xavier to build a sacristy, but the funds were always needed elsewhere. It seemed fitting that this donation be used to fulfill that dream. The sacristy would stand for generations yet unborn as a memorial. Satisfied, he bid the priest farewell, and Jan left the village of his beginnings and returned home.

Marquis de Lafayette

*T*he examination of freedom was the purpose of this life experience. Events during the eighteenth century drew this personality to the Auvergne region of France. Words cannot express what a delight this man was as a child. He was intelligent, curious, precocious, impetuous, and a whirlwind of energy. What a challenge for any parent to raise! What a trial he must have been to his nursemaids!

His father was killed at the Battle of Minden in 1759, and he became an orphan at thirteen when his mother and grandfather died in 1770. Tragedy didn't prevent this personality from observing, examining and experiencing the value of being "self-educated" and remain a "student" throughout his life. He achieved what he did because he allowed life to direct him and personal experience to educate him. And, by acting as an example, he inspired thousands.

The hoofbeats of approaching horses sent him flying to the window.

"Come, Genevieve. Come see."

It would serve no purpose to attempt to distract the boy; he would remain at the window until his curiosity was satisfied. Sighing at the delay, she joined him

He pummeled his nursemaid with questions, especially about the horses. What manner of horse were they? Did they have names? Do horses sleep? If she could not answer the question he would want to know why she did not know. When at last the carriage was emptied of its passengers and was driven off, she successfully coaxed him into his attire.

Stopping him at the top of the staircase, she cautioned, "What did Grand-pere tell you?"

With a great show of exasperation, he responded, "That I must walk down the staircase, no running, no jumping. I do remember."

He refused to take her hand. He was no longer a little boy! But he did as his grandfather had instructed – he walked most dramatically down the staircase.

The guests would visit for two days, Grand-pere had explained as he instructed the boy on his behavior. Upon their arrival, he would join them in the drawing room, welcoming them to their home. He was to remember his manners and, most importantly, ask no questions!

Already he could hear voices and he wanted to break into a run. Oh, it was so difficult to walk slowly, so much easier to rush when there was so much to see.

His eyes roved quickly around the room, taking in everything. A man as tall as Grand-pere stood by the glass doors beside another, shorter and not so old. He was uniformed with many ribbons of honor. The two ladies by the piano were speaking with his grandfather.

About to rush in, a sharp glance from Grand-pere reminded him that he was to approach respectfully. As he was introduced, he gave his little bow as he had been shown and told each he was honored to receive them in his home. His mother's arrival ended his visit. She kissed his cheek and told him it was time for him to see to his riding lesson.

Glancing to Grand-pere he hoped for a reprieve, that he might be allowed to linger a bit longer. But, no, he received only a nod of the head telling him he must do as his mother said.

The next two days were very difficult for the boy because his mother and grandfather were both busy with the guests, leaving him in the care of Genevieve. He liked her well enough but she was always fussing, always telling him what he must do or what he must not do. She neither rode nor played soldier, his two favorite activities.

It was too much to ask that he contain himself for two entire days, and by the afternoon of the second day – as Genevieve sat reading in the gardens – he wandered off. The stables were always great fun and he first went there, plying the stable hands with questions. He made the rounds, investigating the barns until he found the latest litter of newborn kittens. They were still too new to play with so he returned them to their hiding place and moved on. The hens pecked at his hands as he reached to see if they sat on any eggs, and he quickly tired of that game. He did not like the pigs after they were grown fat; too often they became mean and the smell was horrid, so he avoided their pen.

Genevieve was calling him. His stomach told him it was time to eat, but looking at his clothing and hands he knew she would make him wash and change clothing first. Realizing she would be angry, he sighed. She had told him to stay in the garden with her, but he had told her many times that there was nothing for him to do there. She should know this by now.

Tears of relief ran down her cheeks as she scolded him for running off. He must not frighten her so, she had said. And his hands were soiled as well as his clothing. He could not eat until he was clean.

"I am sorry, Genevieve," he said as he saw her tears. "I try very hard to be good."

The next day the visitors were gone. Grand-pere had promised him a long ride, and now they were off – just the two of them. He asked if those men had known his father, did they fight with him? No, he learned, they did not fight beside his father, but they had known what a great man he was. He begged again for the story of his father and he rode quietly as it was told once again. It had been told to him repeatedly, and although he knew it word for word he never wearied of its telling. It was all he had of his father; he had no memory of him for the boy had been only two when his father had died in battle.

"Grand-pere, I have a wonderful idea. You and I shall ride to the end of the world."

"We cannot do so for we would come first to a great ocean."

"Does the ocean go on forever?"

"No, across the ocean is another country."

"When I am a man I shall ride to the ocean and then sail to that country. Will you go with me, Grand-pere?"

Laughing, his grandfather responded, "Oh, mon cher jeune homme, I would go with you anywhere."

Indeed he would have. But it was never to happen. His beloved Grand-pere and his dear Mere both died when he was eleven years of age. With their deaths, he inherited both fortune and title. Well-educated and wealthy, he could have chosen to live a life of privilege, but court life failed to capture his interest. The fawning, taking care to use the proper fork: all seemed ludicrous and unimportant. He chose instead to follow his forbearers' example of a military career. He entered the French Army at the age of fourteen and at age nineteen he was captain of Dragoons when the British colonies in America proclaimed their independence.

He seized the opportunity he envisioned presented by the Revolutionary War in America. As he had once suggested to his grandfather, he did ride to the ocean and sail to that country. In so doing, the Marquis de Lafayette embarked on a career of service well recorded by many historians.

Teacher Madellayne

This personality's love of freedom, education, and America caused it to incarnate in rural Tennessee during the late nineteenth century. Once again we find this personality examining life: the reasons we love, laugh and experience sorrow. She believed there had to be a purpose for it all. No, she never came to any moment of divine revelation. None of us ever do; rather, it comes to us quietly, a moment here and a moment there. But first we must ask the question: what is the purpose for it all?

Once again this personality was too busy being itself and expressing itself to see how she profoundly affected the lives of others. Nor could she see her own goodness and the unconditional love she offered. She never realized what a loving person she was because she entertained unkind thoughts from time to time.

Her father ran the small mercantile that his father had established. Somehow her grandparents had managed to support their families during the Civil War. Her father had been too young to fight but his oldest brother became one of the casualties. For both her parents, the memories of their youth all involved the war, the hardships and sorrows of men who never returned, but neither cared to waste much time in reminiscence, preferring instead to look ahead. Perhaps the future would bring their children prosperity if they worked for it.

Her brother was teasing her again. He had pulled the ribbons from her pigtails and now ran laughing, holding up the ribbons as a prize he had won. How could she go to school looking like such a ragamuffin? Everyone would laugh at her.

"Meanie," she called, sticking out her tongue at him.

Just then their mother came out the door, stood with her hands on her hips and ordered Will to bring her the ribbons. With a sigh, she re-tied a bow on the end of each braid before sending them off.

Will didn't care much for school – one time, he had snuck off and gone fishing. When Pa caught him, he got a real thrashing. Since she started he didn't dare try sneaking off 'cause he knew she'd tell on him. Now, walking to school, she was still so mad at him that she wished he would, just so she could tell Pa. Serve him right if Pa would take him out back again, she thought.

Growing up, Will continued to torment her. When she started sprouting up, he said she looked like a corn stalk; then later it became beanpole, which he eventually shortened to pole.

"Hey, Pole, what's the matter with your hair?" was his latest taunt.

He knew it would send her running to look in the mirror. Her curly hair defied remaining neatly braided, with a curl popping free here and then another there as though it had a mind of its own. She begged her mother to chop it off but was told that would not be ladylike. Ladylike, indeed! How could she look like a lady when her hair looked like a bee's nest?

School was her reprieve; there, she felt as though she was in another world. Learning about the other states, reading stories and poetry, she could lose herself. She didn't especially care for arithmetic; she had to work at getting a good grade. Nevertheless, everyone knew she was the smartest pupil. She was the one awarded the blue ribbon every year at the school picnic.

By the time she completed her schooling, Will was already working at the store with their Pa. Things were somewhat better and they talked of maybe building a bigger store. A few more people had moved into town and with the farmers doing better, business should continue to pick up. With a bigger store they could offer more; already women were asking for buttons and yarn goods. Who knew what was next?

Will's future was already determined, but she wondered about her own. She remained without a suitor even though many of the girls her age were married or at least engaged.

"I'm not all that ugly!" she would think to herself as she looked in the mirror. But, in all honesty, she knew she was too tall to be considered attractive. To make matters even worse, she had no hips! "What use is a trim waist if my skirts just hang from there instead of curving over hips?" she thought.

"Oh, vanity, quit your torment. I am what God made me. Looking in the mirror will never change what I see."

After Sunday church services, she lingered in conversation with her former schoolteacher, Miss Lansing. Her parents were about to leave, and they suggested she join them for Sunday dinner. Quickly Will offered the lady his arm.

Walking behind them, she wondered where he got his manners from all of a sudden! He never even had her as his teacher in school; he was gone by the time she came, so why's he trying to impress her so? Noting Will's face as he looked at Miss Hastings, she giggled to herself. Will's smitten, she realized. And by the schoolteacher! No more sneaking off to go fishing for you, Will, she thought with satisfaction.

Her mother set a fine table. A white embroidered tablecloth, the good dishes and a vase of fresh flowers in the center made it feel like a celebration. On Sunday they always had her special fried chicken with mashed potatoes, gravy and biscuits still hot from the oven. She was enjoying herself watching Will. He looked so awkward trying to remember proper etiquette as the food was passed around the table.

Miss Hastings was pretending she didn't notice his attentions as the pie was served for dessert. Pausing with her fork in midair, she rolled her eyes to the heavens before heaping praise upon her mother. She even asked how to make such a wonderful crust! Oh, poor Will is in such agony it's simply delightful.

Her mother refused Miss Hasting's offer to help clean up, insisting the two young women enjoy the afternoon. They went for a stroll around the yard, chatting not as teacher and pupil but as two young women. How she missed the stimulation of the classroom! As they continued to talk about books and authors, Miss Hastings suddenly stopped.

"Madellayne, you should be a teacher. You love learning, you love discussing things that really matter, you should be doing something with your education."

"Me? A teacher? I don't know, I've never even thought about it. I wouldn't know what to do."

But her mind was racing with the idea. She could see herself in front of a schoolroom, the pupils all sitting at their desks looking at her. If only I could, she thought.

"If you want, I'll help you. We can make some inquiries," Miss Hastings offered.

Walking back to the house, Madellayne said, "Will likes you."

"I know; I like him too. I just don't want him to know it this soon."

The remainder of the summer flew by. With her parents' blessings Madellayne obtained a teaching position. Dresses were made as well as a new coat. She needed appropriate shoes and hat so Miss Hastings took her shopping. With each new article, she added to the trunk that she would take with her, until no room remained.

Her parents worried where she would live. Leaving Will to watch the store, they journeyed to Memphis with her to locate suitable lodging. The rooming houses the school had suggested were unacceptable; the rooms were too small and were not in what appeared to be a good section of town. Madellayne was growing concerned as her father turned the buggy around. Passing a church, her father stopped, got down and went inside. When he returned he told them the minister inside had given him the name of one of his members who may still have a room available.

It was in a nicer part of town, where the homes were larger with lawns well cared for. The owner showed them the room upon learning the minister had suggested her to them. Spacious, two windows hung with white lace curtains, the furnishings appeared elegant to Madellayne. Revealing she was a schoolteacher secured the room for her.

"I allow only women of quality into my home," the landlady informed them as the amount for both room and board was agreed upon. Her father paid for her first month while they set the date for her to move in.

She thought her parents would never leave. She wanted so to unpack by herself, to have the enjoyment of leisurely arranging her things. But her mother lingered, reminding her again of everything she had been told at least a dozen times already. Don't forget church now that you live alone; be sure and save some money each month; don't speak to strange men until properly introduced; on and on the list went. She knew her mother loved her and that it was difficult for her to say goodbye.

Finally she walked to her mother, placed an arm around her shoulder and while walking her gently to the door said, "You and father have taught me well, I appreciate everything you've done for me. I promise to write often and come home every chance I can."

Madellayne had little time to be lonely. There was more to teaching than merely standing in front of the room, she quickly discovered. She read the textbooks, planning the lessons and quizzes. There were always papers to be corrected; she never came home empty-handed. When everything else was done, she read the books she borrowed from the school on how to conduct a proper classroom. If she was going to be a teacher, she was determined to give her pupils the best education she could.

Her landlady was so different from her mother. She really did not care much for the woman; she appeared so dissatisfied with life. The landlady did get along well with the other boarder, though. Both were widows and they spent a great deal of time together talking about how their lives once were. Sighing, Madellayne would remind herself that at least they kept to themselves. She need only share mealtimes with them.

Before she knew it spring had arrived. What a lovely morning, she thought – too nice to waste indoors. Seeing her landlady outside working in her flower gardens, she suddenly wanted to dig in the soil so she went out and offered to help. Oh my Lord have mercy, did that woman become angry! "I never implied that she required my help," Madellayne thought as she flew through the house in retreat.

"Please, Madellayne. I apologize for Mother's rude behavior. She's always been so proud of her flowers, even I'm not allowed to help her," Rose, the landlady's daughter, told her.

The wrong person was apologizing, Madellayne believed. However standing there, looking at Rose, she became aware of a lonely young woman alone in the house with those two widows all day long. Madellayne couldn't recall ever seeing her go out by herself, nor did it appear that she had many callers. Smiling, Madellayne responded that everyone is out of sorts at times, and Rose should not be concerned.

She started spending her evenings in the parlor with Rose; she could as easily do her work there while Rose did her stitching. She was astonished to discover that Rose made all their clothing, even their hats. One morning, Rose had taken her to the attic, revealing where she got her supplies. It was a veritable treasure-trove of hats and clothing waiting to be reused. It was as organized as Madellayne's father's store with hats lined up neatly on shelves, and dresses in trunks labeled according to season. When Rose showed her the boxes of ribbons and the buttons strung on thread, Madellayne remarked that she should go into business. She simply could not believe her eyes.

It felt strange to wake up in her old bed. School was out for the summer and her father had driven into Memphis to bring her home. She had made arrangements to secure the room for her return in the fall. Half the rent since she would not be eating there nor causing her any other work, her landlady had said. It seemed a bit much; she would earn nothing over the summer. However, she wanted to keep the room, to enjoy the summer without worrying where she would live again come fall. Besides, she would not have to take all her things home with her. Her winter clothing, books, anything she would not need over the summer she left unpacked, waiting for her return.

Will and Etta – Henrietta really – were now courting. Etta insisted Madellayne no longer call her Miss Hastings.

"I'm not your teacher anymore, Madellayne," she laughed. "Call me Etta. Everyone else in your family does. I've always wanted to call you Mattie. Would you mind?"

Where did the summer go? she wondered as she packed for her return to Memphis. However, already her thoughts were turning to her classroom, wondering about the new pupils she would have while realizing she would miss the ones who moved to a different room this year. She had the first, second and third grades in her room while the fourth, fifth and sixth were in another. A new teacher would be in charge of the seventh and eighth graders and she wondered who had been hired. This would be her last evening at home until Christmas and her mother had invited Etta to join them for supper.

As they sat enjoying the dessert, Will rose ceremoniously and cleared his throat. She knew! She just knew! He'd finally got up the nerve to ask Etta to marry him.

"I'm not very good at this sort of thing, but Etta and I are going to be married."

"At Christmas," Etta added for him, beaming. "Mattie, you will be my maid-of-honor, won't you?"

It made her last evening easier. Everyone was busy talking about the wedding and where they would live. Etta planned to continue teaching till the end of the school year and suggested Madellayne consider applying for her position for the following year.

Rose welcomed Mattie back; she had missed her company over the summer she told Madellayne. Her room was aired out and ready for her with fresh linens. The two chatted as Madellayne unpacked and readied for the first day of school.

The new teacher was Mr. Orrington. Within a few days she discovered he was single and new to town. He had previously taught at a country school near his parent's farm for a number of years; when they died there was nothing to keep him there, so he sold the farm and looked for another teaching position. He believed his move to Memphis would offer him an opportunity to meet new people as well as increase his income. Madellayne liked the man; his easy sense of humor reminded her of Will.

Leaving church the following Sunday, she was surprised to see him standing at the bottom of the church steps as though waiting for her. He had noticed her in church sitting up front, he told her. They stood talking for a few minutes until Madellayne became aware of Rose and her mother waiting for her. Politely, she introduced them and they welcomed him to their church. She felt he had more he wished to say had they not been interrupted, however when the silence grew awkward she told him to enjoy his day and walked off with the two ladies.

He was not a shy man. He wasted no time in making it perfectly clear to Madellayne that he was interested in her and wished to come calling. Their problem was their teaching positions, members of the opposite sex were not to fraternize; they were to maintain only a professional relationship. It is amazing how many times people can accidentally encounter one another at restaurants, concerts and church socials.

She did not allow him to call upon her at home for she didn't feel she could fully trust her landlady. Not that the woman would do anything malicious intentionally, it was simply that she often spoke without thinking first. She did tell Rose, though, for they had always shared confidences. As with any woman falling in love, she needed to talk to another woman.

As Christmas approached Madellayne prepared to return home for the holiday and the wedding. She had written to her family, telling them all about Mr. Orrington. Etta read between the lines and suggested they invite the man to visit while Mattie was home. When her father arrived with the buggy, she loaded him down with the gifts she had for everyone. She needed to share her happiness even though she knew she had been too free with her money.

Will and Etta were married the afternoon of Christmas Eve Day, and Mr. Orrington arrived in time to congratulate the new couple. The mothers of Etta's students feted them with a small celebration, and Etta cried as she opened the gift from her dear pupils.

With Will and Etta away on their honeymoon, Mr. Orrington was made comfortable in Will's old room. Christmas morning they rose early and exchanged gifts before church. As Madellayne opened the gift he had given her, she looked at him in confusion. It was a key, a rusty old key. She didn't understand and had no idea what to say. He sat looking at her, his eyes dancing with laughter.

"You stole my heart, Madellayne. I thought I could just as well give you the key. With, I pray, the consent of your mother and father I am asking you to become my wife," he said as he knelt down beside her in front of the Christmas tree. Reaching into his pocket, he pulled out a small box and handed it to her. Through her tears she saw the ring. All she could do was nod her assent; she was too happy to speak.

"A June bride. I'm going to be a June bride. Mrs. Alan Orrington, doesn't that sound just delicious? Rosie, oh Rosie, promise me you'll be my maid-of-honor," she exclaimed upon her return to Memphis. Oh, the shopping she must do. Would Rose sew her dress? "Promise you'll shop with me for shoes and all the other million things I'll need," she said as she danced around her room.

So much to do in so little time! With everything else, Rose was opening her own dress shop. Madellayne could not believe her ears when Rose told her. How she ever got her mother to agree was beyond her, but at last Rose would have a life of her own. Mattie wanted to give her friend a special gift and had searched her mind for an idea, discarding everything she could think of. When she told Alan of her predicament, he suggested a sign.

Bless that dear man's heart, what a wonderful idea. He even found someone to paint it for her, right across the front window. Rose loved it. Her shop was wonderful. She had turned that filthy room into a lovely display of hats and dresses. No need to worry about her – she'll do fine if her mother leaves her alone.

Madellayne was married in the same church she had been baptized in. Alan arrived two days early and stayed with Will and Etta. As they spoke their vows, Etta and Rose stood beside her and Will and Alan's best friend stood with Alan. At times she thought she might explode with happiness; it kept filling her insides and spilling over.

She no longer taught; husband and wife could not teach in the same school. She could have looked for a position in a different school, however Alan said no. He bought a house with three bedrooms, one for them, one for the boys and one for the girls, he explained. She had her hands full turning it into a home and, God willing, by this time next year they would be parents. They both wanted children, lots of children.

Before their first wedding anniversary, their daughter was born. They named her Alexis – it was Rose's middle name and Madellayne had always liked the sound of it. Their precious Alexis was baptized in Memphis, and Rose was delighted when asked to be godmother.

A baby for each year, they laughed. School was out for the summer and they were visiting her parents. Etta was as big as a house again, expecting their third baby. Madellayne had just discovered she was pregnant again with their third. Pa and Will were finally building the new store they had talked about for years and they kept Alan busy the entire time they were there. Will said they needed the bigger store just so he could lay in enough supplies to keep his own brood going. Etta laughed in agreement, adding they felt it was up to them to keep the one-room schoolhouse filled with little ones.

It was a struggle, though. The little ones kept her busy from morning till night. Each year she planted a vegetable garden at the back of their yard; it helped some, but money was always tight. They had a small nest egg Alan had started when he sold his parents' farm, and drew upon it only in emergencies. Her parents were always more than generous, loading them with supplies from the store each time they left for home. Clothing was never a problem. Rose remembered every birthday and always had a gift for each of them at Christmas. She would sew a new outfit for each one, but the dress for Alexis was always special. Then, too, clothes were handed back and forth with Will and Etta, whoever they fit this year wore them. Same with shoes: they were worn and resoled until there was nothing left to repair.

She had given herself the afternoon off. The children were in school (all seven of them), her chores were done, and she suddenly longed to spend an afternoon chatting with Rose. She had been in a new location for several years now – this place was larger and Madellayne loved to look at the lovely things she displayed for sale.

Rose was so full of news she had instructed her employee to watch the shop while she and Madellayne went for a cup of coffee. She had been so busy she hadn't had time to drop by and tell Madellayne, but she had bought the shop. "Can you believe it? I am now a proud property owner!"

Then leaning across the table she whispered, "And that's not all I have to tell you."

When Alan arrived home from school she couldn't wait to tell him.

"Our Rosie has met a man. Oh, Alan, you should see her. She's really fallen for him. I know exactly what the man looks like and I haven't even met him. I'm so happy for her, Alan. I just hope he's worthy of her."

"I wonder what her mother thinks of all this," Alan said with the look of the devil in his eye.

"I know. I asked her about that and she said her mother didn't know yet. I give her credit; she's got gumption. I don't know if I could stand up to that woman if she were my mother, God forbid."

"Mattie, Mattie," he laughed, "You would have put that woman in her grave years ago. Remember how we snuck around when we were courting? Or did you forget that already now that you're an old married woman?"

"Married, yes. Old, no," she laughed as she pretended to slap him.

School would be letting out soon for the summer. Saturday the children helped with the chores, the girls inside with Madellayne and the boys outside with their father. This was her favorite day of the week, her family all at home. Even when they were squabbling with one another she loved it; as she listened to their disputes she would remember Will and herself at the same age. It's all part of love, she knew. They just don't know that yet.

As always, they filled a pew in church. Madellayne would remark that it was a good thing she had no hips – if she did, someone would have to sit in another pew. She glanced around for Rose. Not here yet? Sleeping a bit late this morning, are we? What've you been up to, my girl, she wondered mischievously.

The minister walked to the front of the church.

"Before we begin this morning, I have an announcement. It saddens me to inform you all that Rose Andrews met with an unfortunate accident yesterday. Please, join me in prayer as we ask God to receive her in His loving arms.

Madellayne slumped forward, her head resting on the pew in front of her, sobbing. Gently, Alan reached for her and pulled her to him. She could not control her sobbing. She buried her face in his suit coat while clutching the lapel. There was no comforting her here. Instructing the children to remain and walk home after church, he helped her to her feet and guided her out of the church.

"It's not fair, Alan. Just when she had everything to live for; I don't understand."

"I know, Mattie. We'll never be able to understand why these things happen. Why does one man die and another live? There seems no reason to it all. All we can do is go on living, doing the best we can with what God gives us. I don't believe He asks for more than that."

"I don't know, Alan. I would like to believe that, but it just seems there should be a reason for it all, some purpose that gives meaning to it all."

Alan went with her to the bank where they were ushered into a private office. Rose's mother was already there and responded stiffly when they greeted her, again expressing their deep sorrow. They had been surprised to learn Rose had left a will. Even more surprised when Madellayne was asked to attend the reading. She glanced around as they walked through the bank, looking for someone who fit the description Rose had given her, but she saw no one. At the funeral she had been too distraught to pay any attention to the other mourners, and she wished she had for she would have liked to have met the man who meant so much to Rose.

Paying his proper regards to the bereaved, the man began reading the will.

"The last will and testament of Rose Andrews," he began.

The mention of her name brought tears again and Madellayne searched in her pocketbook for a hanky. It was one Rose had given her for her birthday, she remembered and she sobbed all the harder.

She heard hardly a word said. All her memories of her friend were there with her, flashing past one after another. Rose apologizing for her mother, laughing while showing the treasure-trove in the attic, proud of her first shop, standing with me when I became Alan's wife, holding Alexis as she was baptized, ever more confident when she moved to the new shop. Rosie exuberant as she described the man she loved.

Alan had said something to her. Startled, she became aware of everyone looking at her. Mrs. Andrews was glaring.

Mrs. Andrews inherited all of Rose's savings; the shop was left to Madellayne along with a letter and instructions to read the letter later.

The woman was not going to give up without a fight; she wanted her daughter's shop. Madellayne was again called to the bank and notified of Mrs. Andrews' intentions.

"At first I felt guilty that Rose had left the shop to me instead of her mother. I even considered signing it over to her. But, then I read the letter she left me. Here, just a moment. I brought it with me, let me read it to you."

My Dear Mattie,

I hope you never read this, that the day never comes that it is given to you for, if it does, it will mean that I am no longer with you.

I am leaving the shop to you in my will. I am sure Mother will have one of her fits. But, I want you to have it.

You are my dearest friend. You encouraged me, inspired me. Mattie, you were always there when I needed you. You honored me by asking me to be your maid-of-honor, I can't tell you what that meant to me. Then you asked me to become godmother to your firstborn, you even gave her my middle name. Oh, Mattie, your children are as dear to me as if we were related by blood.

But, it is more than that. Family does not always love and honor each other, they do not always encourage and inspire. Sometimes they discourage and scoff at a person's aspirations. You never did. I love you for that. I can never tell you how much; instead I can only leave you my shop to express how I feel.

Remember me in your prayers, my beloved Mattie,

Your Rosie forever!

To reread the letter once again reopened that old wound and she felt the sorrow return. She was grateful she was allowed a few moments to regain her composure.

"I regret it was necessary to call you down here, Mrs. Orrington. I realize it was difficult for you to share your friend's letter, it was written to you and was intended as a personal letter between friends. I don't feel Mrs. Andrews has any legal basis for her claim and I intend to advise her accordingly. However, my advice to you is to keep that letter."

That was the end of it. Madellayne was informed that Mrs. Andrews decided to honor her daughter's wishes. She often wondered what she would have done had it been otherwise. Would she have fought for the shop? She liked to believe that she would have; she felt that is what Rose would have wanted her to do. But she admitted to Alan that she didn't know for certain. It didn't seem right to fight over a person's belongings after they were dead, it just didn't seem very respectful.

Mattie and Alan remained in Memphis: she ran the shop while he continued teaching. Even though she hired a seamstress and retained the clerk, the shop turned a profit. The extra money was a blessing; it helped see three of their children through college. Once they were grown they began to scatter, establishing lives for themselves in other states, and giving Alan and her reasons to travel. Their house was too small to hold the brood they had hatched, so holidays were always at one of their children's homes. No table was big enough to seat all the grandchildren – they would eat on stairways or wherever they could find to sit. They had wanted children – lots of children. And they had them, loving every minute.

Summary

Other than what you have personally experienced, what do you really know for sure? How do you know you're accurate – or are you simply fooling yourself?

The following apocryphal story clearly illustrates that people really don't know what is good or bad for them.

A farmer in India had a very fine horse, but the horse ran away. The farmer's neighbor came to express his sympathy at the farmer's loss. The old farmer's comment was, "Who knows what is bad or what is good?"

The next day the horse came back, bringing with him a herd of wild horses. The neighbor immediately came over to express his joy at the farmer's good fortune. The old farmer again said, "Who knows what is bad and what is good?"

The next day the farmer's son, while trying to ride one of the wild horses, was thrown off and broke his leg. The neighbor came to offer his sympathy. The old farmer once again said, "Who knows what is bad and what is good?"

The next day the army came to the farm looking for recruits. They couldn't take the farmer's son because of his broken leg. The neighbor came over to congratulate the old farmer on the good luck that his son would still be with him. But all the old farmer would say was, "Who knows what is bad and what is good?"

The three life experiences of Angel Nan are additional examples that "you only know for sure what you have experienced". Therefore it is impossible to determine what you are capable of doing until you actually experience it yourself. Knowing that is one thing; personally experiencing it is something entirely different.

As Jan van den Holst, this angel examined the liberating effect of education. At first, Jan took his education for granted. Later on he questioned its value, as well as himself, for he was unable to see how he could apply it. His questions and examinations assisted him in learning how to effectively apply his education. Finally, by example, he showed the liberating qualities of education to many other people.

The incarnation as Lafayette provided this personality with an opportunity to observe, examine, and experience the value of being self-educated. He took advantage of opportunity. What he experienced inspired and motivated Lafayette to remain a student throughout his life, as opposed to seeing himself as an authority. He achieved that understanding by allowing life to direct him and personal experience to educate him. It was his observations, examinations, and experiences that allowed him to achieve so much. Acting as an example, he inspired thousands in the American colonies, France, and other countries.

The incarnation as Madellayne was basically the opposite of Jan's: it allowed the personality to experience what it was like to be a schoolteacher. Identifying and seeing herself as a schoolteacher caused Mattie to experience a different life than the one Jan had experienced. As a schoolteacher, Mattie also experienced the restrictions of that role.

In all three incarnations, the personality was too busy expressing itself and investigating education from different perspectives to see how profoundly it affected the lives of others. Rose's letter to Madellayne allowed her to experience that fact. It is impossible for people to see themselves as they really are for they cannot observe themselves. Fortunately, they are able to see themselves in others because everyone acts as a mirror.

The next time you see something that you dislike in another, ask yourself, "Is it possible that I do that too?" Ask your friends and relatives if you're like that too. When you see something that you like in another, try to observe that quality in yourself. Here again your friends and relatives can help you to see qualities in yourself that you might not be aware of. Is it possible that the qualities and characteristics that you liked and disliked in of each of these angels are contained within you?

Chapter Nine: Angel Sheryl

Marcus Aurelius

The following life experience of this angel takes place in Rome during the second century when Antoninus Pius is Emperor of Rome. Contained within this life experience are numerous examples demonstrating how the personality ensures that it will achieve its objective and that destiny will not be denied.

Every life experience is ruled by destiny as opposed to chance, accident, or coincidence. It is the relationship between life, the personality and the personality's vehicle that determines the destiny of every life experience. Therefore it wasn't by chance or accident that this individual would be taught how to think as opposed to what to think. Everything life and the personality exposed him to ensured that this contemplative and complex individual would become Emperor of Rome.

His father had died when he was yet a boy. Trying to remember all his father had taught him, he now became his younger brother's comforter. When he would go to his mother, too often he found her weeping and unable to speak with him; thus he learned to attend to his brother without the guidance of mother or father.

The household had fallen into disrepair, although not from lack of means, for they were among the nobles. Without the presence of his father much was left undone, and he recalled his father once telling him that a man must make known what is acceptable to him and what is not. He knew his father would not have allowed the neglect of the house staff. It seemed to him that his mother was blinded to the work now left undone.

Oh, that he were of an age when they must heed his words! He longed for their home to be restored to tidiness. His brother seemed not disturbed that he was no longer required to maintain order with his possessions and had become most careless. Reminding his brother that his behavior was unbecoming was to no avail, for their father no longer was with them to render his disapproval.

He welcomed the visit of his grandfather. When he had seen the disorder he chastised the house staff, threatening each with a beating if their work any longer remained undone. They had quickly hastened to their chores, and it gave him much pleasure to see them now rushing lest some task be left undone. He noted how they watched for his grandfather's return, aware that he had gone to speak with his mother.

From that day on it was never known when his grandfather would appear. The house staff, fearing his words, no longer left chores waiting. His brother however had discovered the ease of allowing the staff to see to his things, and his grandfather remained undisturbed by this.

One year past the death of their father, their mother again busied herself with her companions. Often many guests called at their home to enjoy lavish festivities such as he had never seen. He and his brother were not allowed to attend, however he would watch from where he could remain unseen until sleep called him to his bed.

Their grandfather attended to their education; tutors employed by their father reported their progress now to him. The tutors dared not refuse him whenever he arrived unannounced and declared a holiday for his grandsons. These outings became most enjoyable and they learned to listen for his approaching footsteps. He would burst into the room and insist it was time for them to learn more of life than what was to be learned indoors. Off he would take them. It was never known where they would go or when they would return – often several days were spent at their grandfather's house before they were brought back to their studies.

Life at Grandfather's house seemed always an adventure. The brothers suspected that their grandfather must know all people for the house was never absent of callers. Here, they were allowed to remain and encouraged to participate. Their grandfather reminded them that soon they must take their places among men. He would not have them ignorant that they be embarrassed by life. He pondered if such were the way his father would have him learn, but there was no one to advise him.

The sister of his father was an infrequent caller at their house. It appeared to him that their mother was annoyed when she did visit and did not linger in conversation with her. She would depart as quickly as the brothers were called to sit and speak with her. They had not seen her for a year or perhaps longer when she again called on them. She had been ill and apologized for her long absence from them. They could no longer be looked upon as children and he noted that she spoke to them now as men. Inquiry was made regarding time spent with their grandfather and the demeanor of his household. It shocked him that a woman should ask of such things and he hesitated to reply. However, his brother was all too willing to respond in great detail, laughing and believing himself to impress her with his tales. Withholding comment, her face darkened as sadness appeared to settle upon her. Remarking that they had been too long apart, she invited them to summer with her and her husband in the hills. They soon would join all the others departing to enjoy the more favorable air found outside the city. It was an escape from the hot, putrid air that anyone with means looked forward to with anticipation each year.

Summer brought back the memory of his father. His brother had been too young when their father had died to recall his nature, but even now he remembered the peaceful order of his father's house. And so it was here in the summer home of his aunt and her husband. It was most spacious for it was not overcrowded with furnishings. The wall adornments blended gracefully with the hues of the walls, as though part of them. The air carried the natural aroma of flowers his aunt placed daily in the vases. All was most pleasing to the senses.

The days settled into an easy regimen. Their uncle began each day with his walk in the hills, and he invited the brothers to accompany him. As they walked they talked of many things, their uncle often drawing their attention to the order of nature. He had known their father and spoke of him as well. To hear again of his father and the manner of his nature had returned an ache to the boy's heart. That his uncle was a man much like his father became apparent to him. He was a man of honor.

They returned to the city. Their mother had also summered elsewhere, thus the house had been unused. Here summer yet lingered for the air hung heavy in the rooms. Walking through the house he observed the changes that had taken place since the death of his father and he noted how crowded the rooms had become. The walls closed in upon him and he sought space outdoors.

Their grandfather was angered. He had arrived late in the evening and his voice carried throughout the house. They knew he was with their mother but their words were unclear. Pondering what had upset him so greatly, the boy fell asleep.

The morning brought their uncle to their house and the brothers were called to his presence. Their mother sat with them, unspeaking, as he told the boys that they were now to live with him. Their mother and grandfather would be welcomed at his house whenever they wished to visit, however not as a disturbance to their education. It was time for them to grow into the men their father would have them become, he told them before instructing them to prepare to leave.

The quiet strength of his uncle greatly impressed the boy. Visitors to his household were learned and conversations lacked the gossip of the city. They sought not the approval of one another but rather the intellect that they respected. He harkened to their words and pondered how they came to know so many things. His only sadness was his brother, for he waited daily for a visit from their grandfather. He did not come.

By their uncle's public declaration they became his adopted sons. Their previous tutors had been dismissed and new tutors employed. No longer were his studies dull; he now found himself introduced to a thinking that challenged him to look at his world anew. When confronted with a problem that defied solution, he would seek out his uncle's counsel. No answers were readily forthcoming, however, for his uncle would insist on a walk as they continued his examination. At first he had found this frustrating and the walk to be a distraction from the issue at hand. With time, his eyes began to open to the answers that lay before him as he learned to challenge all his teachings.

It troubled him that his brother did not also welcome these new challenges but instead grew more sullen. The closeness they had shared in their youth was gone, and any attempts to speak of this to his brother were met with resentment. Unable to understand this response, he persisted, until one evening his brother became angered with him. In astonishment, he watched his brother storm around the room, raging at whatever lay before him. All the love he held in his heart poured forth as he attempted to reason with him. It served only to focus his brother's rage upon him. He would raise no hand to his brother, and sadly he left the room, unable to see what he had done to incur such vile wrath.

Several days passed before he had the opportunity to speak at length with his uncle. It disturbed him that he no longer could enjoy his brother's companionship, he told him, and continued by relating their encounter. As before, they walked the city, looking for their answer. His uncle pointed to those who toiled, asking him if he felt they enjoyed their life. "Do they toil for love of the labor or do they toil for lack of faculties to secure a position less laborious? If they are unable to do any other, should men of good heart hold them in lower esteem?" he was asked. He was unable to respond. Without further conversation, the two returned home.

It was a problem that remained with him; one that prompted his continued examination of his response to life. It was as a revelation when he finally recognized his brother's anger was not for him but rather for his own limitations of thought. He could not fault his brother for possessing an inability to advance his learning, he discovered. Could he then fault any man? Was there ever reason to fault any man?

His uncle had declared him to be his heir; upon his death, he inherited the throne and became Roman Emperor. That Marcus Aurelius insisted the power be shared with his brother evidences the answers he found within his own heart and the esteem he had for his brother. The benevolent society he attempted to leave as a legacy for all Romans evidences the esteem he had for all men.

John Burgoyne

Events during the eighteenth century provided this personality numerous opportunities to experience nurture from different angles. The British are known for their reserve; the reserve they demonstrate serves to protect them from emotional entanglements. Upon seeing how personal relationships interfered with his sense of well-being, this personality became a master at the art of reserve. Eventually he realized that neither he nor anyone else can escape their emotional nature.

Never completely comfortable with himself, he was driven all his life to succeed. The life experience of this individual is well documented from 1740, when he joined the British Army, until he was buried in Westminster Abbey in 1792.

In the early eighteenth century this personality was attracted to again enter into the physical realm. He was born into neither royalty nor wealth; rather, the family name was recognized for its many generations of service that included military officers, church officials and politicians. With limited means, the family maintained a dignified lifestyle as well as a degree of influence.

He was a frail child. Easily made ill, he spent much time indoors where his concerned mother could oversee him at all times. She limited his activities for fear that he over-tax his health.

His father would have preferred that his only son be robust, and therefore ignored the child. As a boy, he learned to sit quietly when in the same room with his father lest he be sent to his room. Yearning for the companionship of his father, the time passed quickly as he watched and listened. Callers to the home were well educated and cultured. Not allowed to participate in the conversations, he observed their refined behavior while fascinated by the many topics discussed. These visitors brought the world beyond his home to him, thus his youth was one of diversity.

His parents were in disagreement over his education. As he listened from the hallway, it became apparent that his mother wished to have him tutored in the home. His father, however, was determined that his education be more rigorous and had made the arrangements for him to attend a school he had selected in nearby London. He had exerted influence to gain admission for the boy there, believing the experience of living in a male-dominated environment to be of greater benefit. There he would live, returning home only at Christmas and for the summer holiday. As always, his father prevailed.

It was with mixed emotions that he said his farewells to his mother and sisters. He was excited about traveling to London as well as attending school. However, he was uncertain about living away from home amid strangers. It would be a long time till Christmas when he could return to the familiarity of home. Deep inside, he wanted to cry as he left but knew such behavior was unacceptable to his father. Bravely, he set his chin and followed his father to the carriage.

Indeed, it was a very long time till Christmas and he was lonely. Within the first few days he wanted to run away to escape his new life. Although he enjoyed the classroom, he learned to dread the world outside. The other boys would taunt him and seemed to derive great pleasure from his misery; his size and appearance made him their favorite object of ridicule. All attempts to avoid his tormentors failed and he would be forced to listen to them laugh at his cowardice.

When at last it was time for the Christmas holiday he eagerly awaited his father's arrival. When it seemed he had waited too long, he went outside to look for the approaching carriage. He became aware he was not alone only when he felt himself shoved to the ground. He could bear it no longer; angered and close to tears he rose and flew at his tormentor. His father arrived in time to witness his son losing the fight. Stepping in to separate them, he instructed his son to wash and straighten his attire that his appearance may be presentable to his mother while he spoke with the headmaster.

His mother gasped when she saw him and he observed his father's quick response that spared him the added embarrassment of suffering his mother's pity. With his father's strong hand on his shoulder, he was guided into the library and motioned to sit. After an agonizing period of silence his father began to actually speak to him. His father told him that as he grew older life would present him with many reasons to fight. His manhood would be challenged over and over. Wise was the man who ignored the taunts of the ignorant while also being able to recognize those causes worthy of his defense, he was told. They spoke at length of the different methods of defending oneself, whether with words, bodily strength or influential intervention. To believe only one would serve all challenges was to remain an ignorant fool, his father continued, bringing their conversation to a close by encouraging him to learn to determine if it was a fight he could win and whether a victory would be worthy of the battle.

It was a conversation that greatly impressed him, probably because it was the first time his father had acknowledged his presence by speaking directly and solely to him. As he matured, he often recalled those words when confronted with the unfamiliar. His education completed, he returned home a young man in conflict with himself and again in need of fatherly advice.

His interests had been drawn to the literary world while at school: the ability to arouse emotions in another through the skillful usage of the vocabulary appealed to him. The stories he had written showed promise, he had been told, and he had planned to pursue a career in literature. His choice of careers had cost him the company of the young lady that had captured his affections: when he told her of his plans for the future, she ridiculed him. It was the life of a pauper he was choosing, she told him. When next he called upon her, her father informed him that his visits were no longer welcomed.

He did not know what to do: abandon his chosen career that he loved, or abandon her who he also loved, for he was unable to see how he could have both. His father listened without interruption. When he finally had no more to say, his father proceeded most cautiously for he was wise enough to recognize his son had grown to maturity. He sympathized with him over the quandary he was confronted with and agreed that there appeared to be no ready answer. Because one is at times too involved with the problem to recognize the solution, it may be of benefit to withdraw oneself, he suggested. Time spent in other fields of endeavor may sometimes prove to be of benefit for they allow one a fresh approach to the problem, he continued. The young man agreed that perhaps that was what he needed – to be preoccupied with other pursuits that this dilemma could not plague his every waking moment. Supper interrupted their discussion and his mother and sisters took up his evening.

The following morning he and his father escaped the women by seeking the out of doors. As they walked, his father said he had pondered the problem at great length before sleeping. He had examined several alternatives before recognizing the advantages offered by time in the military. It would be time that demanded his total attention as well as providing him an opportunity to form strong associations. If he agreed, his father was willing to make inquiries to secure a position for him. Without hesitation, the young man asked his father to do so for he welcomed the escape it offered.

Focusing all his energies on succeeding as a young officer in the British army, he rose in rank. Military campaigns in Portugal brought him into prominence and he was considered to be a man of promise. At the urging of respected friends and associates he entered into the political arena, serving several years in Parliament. The conflict with the American colonies prompted him to return to military service. Believing he could bring the colonial resistance to an end by isolating New England from the other colonies, he was given command of an army. Although it would have been a victory worthy of the battle, General John Burgoyne recognized too late that it was a battle he could not win: his army was poorly equipped and wholly untrained for the frontier fighting they encountered. After suffering heavy losses, he was forced to surrender his army. He returned to England in disgrace and after parliamentary inquiry into his conduct of the war, resigned his rank.

Although later reinstated for a brief command in Ireland, he retired from military life. While serving in Parliament, he had returned to writing and his trunks contained several unfinished manuscripts. Rereading them, he burned those that displayed his early attempts at expression. Only two he retained, determined to rewrite and complete them. His remaining years brought him recognition for his literary abilities and he enjoyed an active social life. However, he never married for he remained forever unsure whether that victory would be worthy of the battle.

Strasbourg Physics Professor

Seeking balance in the field of academia, this personality is attracted to diverse Strasbourg, Germany during the late nineteenth century. Strasbourg means "City of Roads," an appropriate name for a city that has spent centuries at the crossroads of Europe. It is also a city that has been passed back and forth numerous times between France and Germany.

Once again this personality examines the influence of nurture from a different perspective. Unlike Marcus Aurelius, he was taught what to think rather than how to think. His needs were simple and he asked very little from life. He enjoyed his work, his associates and his home. These were enough for this shy and naïve individual until his life was shattered. Culture, academia and the absence of practical experience left him unprepared for what was about to occur in 1914.

He was born the same year Strasbourg again became part of Germany. His father always told him he shared his birthday with the rebirth of their city, for their family ancestry was German.

The family enjoyed a good life. His father was a professor in the mathematics department at the University of Strasbourg and was highly respected. Their home encouraged cultivation of the mind as well as appreciation for the arts. Family friends were those associated with his father at the University; thus the environment was rich in diversity.

From the time he learned to walk, he adored his Papa. Perhaps because he was the youngest of their sons or perhaps because he was born so much later than the others his Papa took time to enjoy this child. His earliest memories were those of sitting on his Papa's knees while he learned to count: eins, zwei, drei . . . , eins, zwei, drei . . . , patiently his Papa would count with him on his fingers. He remembered how proud Papa was the day he met him at the door, counting alone without error. Papa lifted him upon his shoulders and they marched around the house, counting loudly as they went from room to room.

By the time his formal education began, his mind was eager to learn. School was easy for him, and he could not understand why some of his classmates had so much difficulty. He often spent much of his time trying to assist his friends with their lessons so they would have more time on the playground. Athletics had not been encouraged at home and he found them to be great fun. When his teacher praised his athletic coordination, he could tell that his Papa was not pleased. When they reached their home the two of them talked about the purpose of education and how easy it is for a man to allow other diversions to interfere. Playtime is time wasted, he had been told, time that can never be recovered.

He sensed he had disappointed Papa. He wanted never again to give him any reason for distress and he made this known by passionately promising to apply all his time to his lessons. Silently, he vowed that he would do nothing that would not make his Papa proud of him. Never again did he wish to see he had given him cause for disappointment.

As had his brothers, he attended the University where their father taught. Excelling in all his studies, he at first decided to emulate his father and enter into the field of mathematics. However, once at the University the science of physics captured his imagination for the countless possibilities it presented. His father expressed his approval, encouraging him to advance his studies that he may apply for a professorship.

He was aware of how proud his father now was. They would end their week by lunching together, discussing their students, current events, and family news. His brothers were all well established, married and had started their families. Not one to pry into another's personal life, his father's desire for additional grandchildren was apparent. He was now thirty years of age and without a social life. His time had been devoted first to his studies and later to his profession. Living at the University, the academic world had become his entire life and he was comfortable only with his associates there.

His father invited him home for Saturday supper. An old friend of his was visiting from Berlin accompanied by his family. They had been friends since their boyhood and had attended the University together, his father explained. His friend had moved to Berlin years before and now taught at the University of Berlin. No excuse was acceptable for not joining them, his father insisted – he wanted his entire family to welcome these visitors.

His mother had cooked all her finest German dishes and they all feasted. He found himself relaxing as he listened to the two old friends reminisce about the days of their youth. After everyone had eaten their fill, they moved to the parlor while the women cleared away and washed the dishes. Later, when they joined the men in the parlor his father turned to his friend's daughter, Hilde, and asked her to play something for them on the piano. Her father had proudly spoken of her musical talent and he now encouraged her to entertain them with her music.

He sat listening to the music that filled the room and recognized the concerto, for it was one of his favorites. And he became aware of the pianist. Throughout supper he had paid no attention to her, for he was preoccupied with his own thoughts; now however, he observed her graceful movements as her fingers moved across the keyboard. When the music ended, he rose with the others and expressed his appreciation for her talent.

Their courtship was long distance for she lived in Berlin with her parents. They began by writing long letters to one another and, when the letters were no longer enough, he traveled by train to see her. Her parents welcomed him into their home; he was their houseguest whenever he visited. It was two years before he found the courage to ask her to become his wife.

He had always been frugal with his money, living a simple life in a single room at the University; he had not needed anything more. With his savings he purchased the house where he would bring his bride. As he stood in the house looking at all the empty space, he wondered how he could afford furniture with the funds he had remaining. He needn't have worried. His parents gave him the bureau he had used since childhood along with the bed they bought as a wedding gift. The dining room was the gift of Hilde's family and her piano was shipped by rail. Along with his desk and a few other items he had purchased for his bachelor's room, the house was quickly transformed into their home.

Where work had once been his entire life, he now discovered himself hurrying home at the end of his day. The aroma of Hilde's cooking filled the air whenever he opened the door and he came to love the smells of his home. Here was where he realized what contentment was, where he came to know another person as well as he knew himself.

Annually they traveled to Berlin to vacation at the home of Hilde's parents. Hilde enjoyed shopping with her mother as well as visiting friends she had known since childhood. Over time, he had come to know several of her father's associates there and he always made time to call upon them. Each year they encouraged him to join them, assuring him that they would recommend him to a position there. Although he admired and respected all of them, after a month in the city he was always ready to return home to Strasbourg.

They remained childless. He was aware of the sadness this was for Hilde, the room she had prepared for the children they anticipated no longer stood open. She had closed the door so she need not look into the unused room. He had made several attempts to speak with her about the possibility of adoption, but she was unwilling to consider the option: it would be admitting her failure to bear a child of their own.

Their marriage became troubled. When his parents died, Hilde began encouraging him to accept a position in Berlin where they would be close to her parents. With the outbreak of the war, she became persistent that they move. He now found himself more undecided than ever. Although he was of German descent he found himself questioning the actions of his government. When German forces moved into France, his loyalties became divided. The city of Strasbourg had once been part of France, and some of its citizens were still French.

His oldest brother lived in Paris along with his family and he feared for them. He wrote to his brother, begging him to return to Strasbourg. Each day he waited for his reply and he became increasingly worried. Hilde reminded him that mail service could no longer be relied upon. She expressed her fear of an invasion by France to recover the territory it had lost. Weeping, she begged him to move to Berlin, where they would be safe.

Safe! His mind seemed to explode. Was anyone safe, anywhere? He felt his brother was not safe in Paris and wanted him to move to Strasbourg. Hilde did not feel safe here and wanted to move to Berlin. *Has the world gone mad?* he wondered. *Who is the enemy? If I am a German and Germany is the enemy of France, does that make me also an enemy of France? If my brother considers himself a French citizen, is Germany his enemy and am I my brother's enemy? What is an enemy?* He could not face the questions that ran wildly through his mind for they were without answer. He bolted from the house. He needed to think, to clear his mind so he could see what to do. He walked the darkened streets for an hour, pondering. Unable to find any answers, he returned home.

As he sat at the breakfast table the following morning, he observed several birds outside their window. Amid the insanity created by man, they still sat upon the branches to greet the day with their songs. Although of different species they shared the same tree; somehow they accepted each other's right to be there. Why cannot man do the same, he questioned? Are we less intelligent than they? Is there something they know that we do not?

He knew he was unable to make any decision about a move to Berlin without first speaking with his brother, and this he told Hilde. He needed to know that they were safe and what their plans were. If he could not rely upon the mail then it was apparent that he must go to Paris. Hilde argued with him, reminding him that as a German citizen he would perhaps find France unwilling to welcome him. He agreed to speak with his brother living in Strasbourg before leaving, and his brother insisted they travel together. Hilde would stay with her sister-in-law until they returned.

They traveled by rail. At the border, the train stopped, everyone was ordered off, and they were informed that they would have to board another train to complete their journey. As their papers were inspected, he found himself becoming fearful. Never had he felt this way before and it angered him to realize that his own country had created an environment of fear. Relief settled over him when the train began to move. Glancing at his brother he recognized the fear had not been his alone.

Looking out the train window he saw what appeared to be smoke in the distance. Remarking to his brother, others began to notice also. Then the train began to slow until finally it came to a stop. Without the noise of the train wheels rolling on the tracks, the sounds of war could be heard. The conductor walked through the train car informing the passengers that it was unsafe to travel any further for there was fighting up ahead. Each could do as they felt best, but he told them he was leaving the train for it was too large a target. There was a small town within walking distance and anyone wishing to do so could join him.

The brothers were uncertain what to do. Leaving the train, they paused and glanced around. They had not left home to end up hiding in some small village, they both agreed. Better they try to follow the tracks that they knew would lead them to their intended destination. They could always seek cover in trees along the way or lie flat along the tracks, if needed.

They set out on foot, at first walking in the fields. After several hours it became too difficult and they moved up onto the tracks, walking between the rails. Coming to a bridge, they hesitated and looked down at the river below. Both were thirsty and in need of water. They decided it was better to drink now, while water was available, for they had no way of knowing when they would find water again that day. They scrambled down the riverbank, looking for an advantageous place to drink from. Several large boulders under the trestle appeared the easiest and each brother headed for one. As he bent over, reaching his hand into the water, he heard a sound he had never heard before. He was choked with fear as he looked up to see the bridge explode. Suddenly it was raining splinters of wood. Turning to head for dry land, he slipped. As he fell backwards, he looked up and saw a railroad tie falling straight towards him. It was the last thing he saw.

Summary

The capacity for feeling, expressing and managing your emotions is far more important than you realize. Emotions are the intelligence's connection to consciousness. Emotion is the language of consciousness just as feelings (the body's senses) are the language of the body, and thought the language of knowledge. Those who ignore, rationalize, or deny their emotions are oblivious of what is going on inside themselves. They are hopelessly naive, hollow, and insensitive. Those who don't express their emotions exist in a black and white world as opposed to living in a world filled with millions of colors. Yes, sometimes it can be difficult to express that you're sad, worried, or upset. When you don't express these emotions then it is just you and your bad feelings. Keeping them locked inside limits your potential and causes you to be ill and unbalanced. Illnesses, diseases, and accidents first manifest on an emotional level.

On the other hand, those who don't use their intellect and the body's senses to keep their emotions in check lose control of their emotions. Those who become overwhelmed by their feelings experience distress, discouragement, and depression. Those who don't control the expression of their emotions vacillate between depression and rage. It's like being on an emotional rollercoaster ride.

You're not alone if you have experienced any emotional difficulty. Prior to the twentieth century many saw thought, emotions and feelings as being the same, and it wasn't until the twentieth century that science started discerning the characteristics, strengths and weaknesses of emotions, feelings, and intellect. It wasn't until the latter part of the twentieth century that science recognized the intelligence of emotion and labeled it Emotional Intelligence (EI). Likewise, Emotional Intelligence Quotient (EQ) is your ability, capacity, or skill to perceive, assess, and manage your emotions. Today, science realizes that EQ is just as important as IQ.

The life experiences of Angel Sheryl are about discernment and shutting down emotionally. This personality has been examining the influence of education (nurture) for a long time. It is important not to judge this angel for it was the personality's intention to experience life void of emotions. I initially experienced some subtle and unexplainable conflict when I examined these incarnations of Angel Sheryl. Eventually I realized what the conflict was: these incarnations were really out of character for this dynamic and vibrant personality. The more I examined what I was experiencing, the more I realized how cleverly disguised this personality was. I'm also aware of the difficulty this personality is currently experiencing today as it works to reconnect with its emotions and to express them so that it may regain its balance. I felt the internal struggle or war between heart (emotions) and intellect (thought) that this personality was experiencing. Life is working with this personality today to complete its review incarnation.

The life experience of Marcus Aurelius reveals how his adult life reflected the influence he received in his youth. His life was influenced by what he personally experienced and by his education. The diversity of both these influences caused him to be a very reflective individual who constantly challenged the findings of his observations and investigations.

John Burgoyne's life was different. His father's inability to express and show his emotions caused John to attempt to please his father but – like his father – John was unable to express his emotions and feelings. Not having received much affection in his youth, he found it almost impossible to understand those who expressed their affection, feelings, and emotions. Although John was very intelligent and well educated, his sheltered life prevented him being exposed to a diversity of experiences. His plan to crush the colonists was logical and intellectually sound, but unfortunately it was destined to fail for it was not based upon personal experience. His failure in defeating the colonists balanced his success at ruling Rome.

The physics professor's absence of diversity and experience was even greater. So much so that it prevented the personality from experiencing anything that was not academic. It took a war to cause him to start questioning things, and his lack of experience brought about his death.

While recording the life experience of Marcus Aurelius, I became aware of the intelligence of this man: in many ways he was as intelligent as Albert Einstein and was way ahead of his time. His adoption by his uncle was extremely beneficial for it afforded him the opportunity to be introduced to philosophical thought. Greek and Roman tutors challenged his intellect while his uncle continued his education by encouraging him to examine his own world.

In today's world, he would be the student easily bored with education because it neither stimulates nor challenges him to stretch beyond himself. His ease of learning made it difficult for Marcus to understand his brother's slow mind. Had the brothers remained under the influence of their grandfather his brother would shave been happier – their grandfather expected little of him, and he was comfortable with that. It would have been a life of the idle rich with too much time wasted on pursuit of pleasure. It was a lifestyle the brothers were encouraged to participate in and would have become theirs also had not their aunt and uncle interceded. Destiny will not be denied!

The recognition that not all men possess abilities to advance their learning awakened his compassionate nature. Marcus was a contemplative and complex man, and he attempted to create a benevolent society that cared for the less fortunate. As with any charitable attempts, his were less than successful and at great cost to the treasury. Although highly intelligent, he did not understand that government cannot assume the responsibilities of the individual. His introduction to philosophical thought and being self-educated remained with him throughout his life. He kept a spiritual diary, "Meditations," where he continued his examination of life.

The incarnation as British General John Burgoyne served as a transition between Marcus and the physics professor. It also ensured that the personality would experience what it is like to be totally shut down emotionally. As a sickly child, John spent most of his time indoors, thereby limiting his ability to interact with other children his age. Although he spent most of his time indoors, he was unable to receive the attention and love he desired from his father. The emotional trauma he experienced when he was sent off to military school caused him to ignore his emotional feelings, and his military training reinforced that behavior.

John had become very adept at expressing his feelings verbally and through the written word. Unfortunately his desire to be a writer was shattered when the woman he loved ridiculed his ambition. That experience and his career in the military further reinforced his decision to ignore his feelings. John's involvement with politics and Parliament provided an outlet for him to verbally express himself. Eventually in the winter of his life he experienced some literary success.

John's desire to win the approval of those who meant something to him (such as his father) caused him to be fearful of failing. When you combine that with his lack of practical experience, it was inevitable that he would fail in both love and war.

As a physics professor, this personality experienced the benefits of an education once again. Unlike other life experiences, the name of this Strasbourg University professor was deliberately withheld. One of the reasons for doing that was to illustrate that it isn't important to know the name of this individual or what he experienced. What is important is seeing how the personality perceived what it experienced and how he responded to its perception. Names also assist thought in falsely identifying itself with life experiences that it never experienced.

Although not a member of the military, this personality experienced the conflicts created when countries are in opposition to one another. He found himself in a difficult position; born into that portion of the world that had once been part of France, it was again under German rule during his lifetime. When Germany became the aggressor, he found himself confronted by questions he was unprepared for.

His life had been one of culture and academia. To attempt travel may appear to have been foolish, however it must be recognized that both he and his brother were extremely naive. Their lives up to that moment had provided them with no relevant experiences to draw upon – in many ways, they had led a sheltered life. He was very shy and inexperienced with women.

His profession was his passion and he allowed it to fill his life until he first heard Hilde play. When he became aware of his attraction to her, he still proceeded very slowly. He was not certain she would accept his marriage proposal and felt very fortunate that she did. His needs were simple; he asked very little from life. He enjoyed his work, his associates and his home, and these were enough for him. But the war was changing everything, and in many ways he felt lost, for his world was falling apart and he could not find the pieces to put it back.

If you don't keep a daily journal or diary, you might want to ask yourself, "Why don't I keep a daily journal?" If you do keep a journal, keep track of unpleasant experiences that keep repeating. Pay particular attention to those that affect you the most. Record your perception of them, how they affected you, and how you responded. Doing this will allow you to see and understand "cause and effect" relationships. You'll also want to keep track of your insights and accomplishments. It is the experiences you gain from doing this that will enable you to change your life without effort, denial, or willpower.

Part III

*Understanding What
Life Is All About*

**Ah, to feel; that's to know
you are alive –
and living is all about
the journey!**

Chapter 10: The Missing Link

*Life is a multidimensional journey that
can only be understood by consciously
experiencing it physically, mentally,
and spiritually.*

– David Zimmer

By now you are probably wondering what the relationship between Part I and Part II is. Part I intellectually challenged and confirmed various perceptions you had about the human experience and the evolution of the personality, and then Part II provided you with a realistic, visual representation of what an evolved personality is capable of achieving. It also provided you with a different view of angels from the one you are accustomed to seeing. The missing link that bridges the gap between Part I and Part II is your own life experience. It is what disagreed, agreed, or a combination of the two with what was presented in Part I. It is what allowed you to see that incarnated angels are more human than angelic. And it is what made you realize that you cannot physically recognize an angel.

The key to understanding life, angels, others, and yourself is contained within you as opposed to outside of you. That is why your life experience is the foundation upon which this book was built. At best all this book can do is make you aware of what lies within you and assist you in deciding what to do with it. You realize that fact by experiencing it, and that's the purpose of the balance of this book.

The human body is a physical, mental, and spiritual vehicle that allows you to experience life from a multidimensional perspective. Whether you are aware of it or not, everything you experience registers upon you physically, mentally and spiritually. The body's senses, feelings, and the brain allow you to be aware of what registers upon you physically. The contents of the brain and thought cause you to see yourself as being a physical entity. However, they prevent you from seeing yourself as a mental and spiritual entity because they are oblivious to everything you experience mentally and spiritually. Only that which is non-physical can see, experience, comprehend, and understand your mental and spiritual nature, as well as your physical nature. Only consciousness is able to achieve that because it is non-physical. Consciousness and a quiet mind allow you to experience and understand your physical, mental and spiritual nature. They do it by providing you with the awareness and the experience you need to interpret and understand your physical, mental, and spiritual nature.

Just as thought is the language of knowledge and feelings are the language of the body's senses, emotions are the language of the personality. **Emotions bridge the vast gap between the physical and the non-physical.** They assist you in being aware of your spiritual nature. Asking yourself, "What is the source of emotions?" assists you in realizing that. Unlike physical eyesight, realization is not a product of the senses. It is non-physical sight, an aspect of your mental nature. Your mental nature interprets and makes sense of your physical and spiritual nature. Consciousness is your mental nature.

Emotions allow the physical body to experience what the personality has experienced and the personality to experience what the physical body has experienced. Like English, French or Spanish, the languages of thought, feelings, and emotions are a vibration. Feelings are a higher vibration than thought, but a lower vibration than emotions. Although thought has knowledge of feelings and emotions it is unable to understand them because thought is unaware of them and cannot experience them. Conscious awareness and experience are the keys to understanding anything, for you can only understand what you have consciously experienced.

Experience and emotion are a common bond that allows you to understand others. Only someone who has suffered the loss of a loved one is able to understand another who has experienced a similar loss. Likewise, an alcoholic is able understand other alcoholics because they share a common experience, whereas someone who isn't an alcoholic doesn't.

Having stated all that, let us see why **you** are the missing link between Parts I and II. You are unique – not duplicated. That means no one else has your personality, your perception, or has experienced everything exactly as you have. You are the only one who has and who ever will make the same journey through life, from one incarnation to the next. That means you are the only one who knows what is best for you, is able to understand you, and is able to resolve the problems you have created. Whether you are able to accept it or not, **you are your own savior!** That is why you are completely and totally responsible for your "self". It is also why it is impossible to delegate, transfer, or entrust that responsibility to another.

If you are unwilling to unconditionally accept responsibility for yourself, then organized religions, governments, self-help experts, gurus, and other "authorities" will be able to convince you that they can be responsible for you. The illusion they create is appealing because you perceive it to be simpler and easier than being responsible for yourself. Furthermore, it provides you with a socially acceptable excuse if that responsibility isn't fulfilled. Whenever you attempt to delegate your responsibility for yourself you agree to participate in the illusion and experience the consequences of it. And those consequences cause you to falsely think and believe that the illusion is real.

Part II helps you see what is contained within you and to see how much you have in common with angels. The more you are able to see yourself in another, the more you understand them as well as yourself, because everyone serves as a mirror. In other words, you can only recognize in another what is contained within you. Whether you are aware of it or not, what you like about another is what is contained within you. What you dislike about another is what you dislike about yourself. What is contained within you caused you to see some of the life experiences of angels in Part II as more interesting or appealing than others.

When you see yourself as being only a physical entity, it causes you to see others as being different from yourself. In turn, that illusion causes you to compare yourself against others, thereby corrupting your perception and perpetuating the illusion. It also inflates your ego and strengthens thought's control over you. All this increases the separation and furthers the distance between you and others.

Seeing yourself as a spiritual entity unites what thought separates. It helps you to see yourself in others and allows you to communicate with others on an emotional rather than an intellectual level. This can be easily seen and experienced if you reread Part II from an emotional rather than intellectual perspective. You do that by attentively looking to see what characteristics, behavior, thoughts, and emotions you have in common with these angels. In other words, you are observing with your heart rather than with your head.

Because emotions are universal they provide a common bond that unites those of different nationalities, religions, race, and social status. If you ignore, deny, and rationalize your emotions you will feel very uncomfortable with anyone who connects emotionally with others. You will also find it extremely difficult to see yourself in others. Although you might not be consciously aware of that, those who are comfortable with dealing with their emotions can easily recognize those who are not.

The flip side of seeing yourself as a spiritual entity also allows you to put yourself in another's shoes. Doing that develops empathy, compassion and understanding. Case in point: Most people don't see that the war in Iraq actually created **more** insurgents and violence. They could see this and understand it if they asked themselves, "How would I react if my religion was being attacked and my country was being invaded?" Did the number of military recruits increase or decrease when Japan attacked Pearl Harbor, or when Germany attacked various parts of Europe during World War II?

Angels serve as role models for those who are able to see what they have in common with them and who are able to put themselves in an angel's shoes.

Keep in mind that once the personality is incarnate it experiences everything its vehicle experiences. The personality is unable to access its memory, so it has no knowledge of the arrangements it made to ensure that it would achieve its objective or what it desired to experience. The knowledge the personality has is now limited to that of its vehicle. Thus, for all practical purposes, the personality and its vehicle are one and the same. Both the personality and its vehicle are completely dependent upon life to provide everything they need to develop their potential. You also know that resistance develops potential and gives it worth. All this suggests that there is a reason for everything you experience. That means:

> Everything you experience is important
> and needed. Nothing you experience is
> ever wasted or for naught.

If this is so, then surely it is wise of you to observe how you perceive each thing you experience, how each experience affects you, and how you react or respond to them?

When I didn't understand the purpose and value of what I was experiencing, I judged, criticized, and tried to escape every experience I saw as being bad, painful, or disastrous. Most of my day was spent resolving problems, rehashing the past, and worrying about tomorrow. Life didn't make any sense to me and I wondered why it was so difficult. I vacillated between coping with life, attacking it, or trying to escape it through alcohol, sex, food, and other activities. In turn, I wondered why I was unable to improve the quality of my life and obtain what I desired.

Eventually I found myself trapped in an extremely unpleasant situation that was impossible to escape. At first I resented having to experience my predicament and tried everything I could to escape it. Eventually I realized that it is impossible for me to escape myself – self went everywhere I went. Seeing that allowed me to unconditionally accept my situation. The moment I did that, I saw my situation and everything else differently. In turn, everything I experienced was different and I observed that I reacted differently to everything I experienced. Over the course of about ten years I found myself in a couple of similar situations. In time I realized that my perception determines what I experience and how it affects me emotionally, mentally, and physically. And that determines how I respond or react. For the first time in my life I realized that I could control what I experience by questioning and challenging my perception. I also realized that my relationship to life helps me do this because everything I experience challenges my perception.

Life and I have a relationship with each other. That relationship makes life totally responsible for providing me with everything I need to develop my potential. Likewise, I am responsible for using what life offers me to develop my potential. Experience has allowed me to see that life always fulfills its part of this relationship but I do not always fulfill my part of our relationship: I vacillate between being responsible and being irresponsible.

> "As human beings, we are endowed with freedom of choice, and we cannot shuffle off our responsibility upon the shoulders of God or nature. We must shoulder it ourselves. It is our responsibility."
>
> *– Arnold J. Toynbee*

Responsibility is a very difficult concept to understand due to its relationship to other things such as discernment. The more discerning you are, the more responsible you are, and the more responsible you are, the more discerning you are. Recognizing the differences between *responsible* and *irresponsible* develops your ability to discern. It also helps you see that irresponsibility is often disguised as responsibility. An example of that is assuming the responsibilities of another. Asking: "Who is responsible?" and "Who caused or created the problem?" helps prevent you from assuming another's responsibility. The life experiences of Angel Lee (Henry of Jamestown, Betsy Ross, and Mary, South Dakota Pioneer) are an examination of "control" as well as discerning the differences between irresponsibility and responsibility.

Life is a series of never-ending choices. The individual you are today, the quality of your life, and everything you have experienced were created by the choices you consciously and unconsciously made. That is a difficult concept to accept because it makes you completely responsible for yourself. The reality of it is humanity's biggest dilemma.

Chapter 11: The Journey

Life is all about the journey! What does that mean? Information is useless until you *understand what it means.* Understanding what it means is a dynamic, interactive and multifaceted process. It involves experiencing, observing, interpreting, and testing your findings in a particular way. All this requires a viable means to discriminate the essential from the nonessential and the integrity to accurately determine the validity of your findings.

Case in point: Man has been attempting to put an end to crime, wars, poverty, and hunger for eons. What has prevented him from being able to solve these problems? Surely it is not the lack of knowledge, because he is drowning in it. Man doesn't understand "the meaning of understanding" and what is causing his problems. His attempts to resolve his problems by focusing on the *effect* rather than the *cause* confirm that fact. He doesn't understand that *causes* don't exist in the observable world because the observable world is a world of *effects*. Consequently, man relies upon force instead of power to resolve his problems. Force comes in many different forms, such as coercion, taxation, laws, and war. The use of force increases man's problems because force consumes tremendous amounts of energy and is extremely costly. Force doesn't resolve anything! In fact, history clearly shows that it has brought about the demise of one great civilization after another.

> "We can't solve problems by using the same kind of thinking we used when we created them."
>
> *– Albert Einstein*

The source of all of Man's problems is thought or some aspect of it. Thought is the source of all humanity's problems. Thought is also a master at creating illusions. The illusions thought creates appear to be real because thought perceives them as being real and the brain cannot distinguish an illusion from reality. It takes something other than the brain, thought, or intellect to see, understand, and destroy the illusions and problems created by thought. Consciousness is able to do that because it is different and in a higher field of existence than thought. That is why it is so important to rely upon consciousness rather than thought.

One of thought's illusions causes you to think and believe that another individual can make you angry, happy or sad. All another can do is make you aware of what is contained within you by doing something that stimulates it. Although thought may have knowledge of that, it is unable to see the anger, happiness or sadness that is contained within you. The following saying assists you in realizing that: "Sticks and stones may break your bones, but words can never hurt you." Does knowing and comprehending that saying prevent words from hurting you? Why?

You have the ability to choose. You can choose to continue doing what man has been doing since he came into existence and expect a different result. Or you can choose to consider the possibility that you don't understand what you're doing or dealing with. The first option is the definition of insanity. The second option is the first step towards understanding. The hardest thing about the second option is acknowledging that you don't know what you think you know and believe.

Is your perception of angels today the same or different than it was prior to reading this book? If your perception of angels changed then you have already acknowledged that your previous perception of angels was inaccurate. Could it be possible that your perception of what life is all about is also inaccurate?

Experience

Every moment of every day you are experiencing something. As long as you are incarnate it is impossible not to always be experiencing something. Life and every aspect of it is an experience, and every experience is multi-faceted. Just because you may not be aware of what you are experiencing doesn't mean that you are not experiencing it.

How would you view and respond to everything you experience if you realized that it actually assists you in obtaining what you desire? Let's see if that statement is as absurd as it appears to be. Do you recall the relationship between life and the personality? Rather than simply reading it or allowing thought or intellect to interpret it, ask yourself, "What does that mean?" If you do this you will see that life is always providing the personality with everything it desires. It is impossible for life not to do that due to its relationship with the personality and the Universal Laws that govern life.

So what does the personality desire? Every personality desires to understand itself and to develop its unlimited potential by experiencing what it is not. Keep in mind the fact that the personality is limited by its vehicle. Did the life experiences of Thomas Beckett and Joan of Arc develop the potential of Angel John's personality? Did the life experiences of Pastor John, William, and Nattie develop the potential of Angel Lyn's personality? Was it what these personalities experienced that assisted them in developing their potential or was it something else? How would you view and react to everything you experience if you realized everything you experience does the following:

- One: it gets your attention. Whether you are aware of it or not, everything you experience registers upon the mind and body. Unfortunately most of the time you're unaware of it because of what you're focused on due to thought. More often than not life has to repeat the experience over and over again to obtain your undivided attention. If that doesn't work it uses accidents, illness, and disasters to get your undivided attention.

- Two: it causes you to pause or stop what you are doing so that you can see and become aware of what you have been doing. Just as you have to get out of the trees to see the forest, you have to stop what you are currently doing to see what you have been doing. What you experience stimulates your body's senses and your emotions. In turn, they get your attention and cause you to stop what you have been doing. If that doesn't do it life uses bad accidents, weather, and disasters to stop you from continuing to do what you have been doing.

- Three: it challenges your perception – what you think, know and believe. More often than not your perception is preventing you from obtaining what you desire. Life helps you achieve your objective by introducing experiences that challenge your perception. Did this book challenge your perception? Was it by chance, accident, or coincidence that you are reading it?

- Four: it tests your values, character, and integrity. You are constantly being tested because you are always experiencing something. It is how life has evolved and will always continue to evolve. It can do no other because of the Universal Laws that govern life. Another word for this testing process is "evolution".

- Five: it assists you in seeing that your perception is preventing you from obtaining what you desire. This is more difficult to see because it takes a while for you to see that what you experience challenges your perception.

In turn, it takes countless challenged perceptions in order for you to see that your perception is preventing you from obtaining what you want.

- Six: it is an opportunity to experience what you desire to understand. This is even more difficult to see than the previous one because it is so transparent and it cannot be done with thought. It requires that you consciously participate in the experience and be observant enough to understand the reason for the experience.

- Seven: it helps you to be what you desire to be. Everything previous to this is assisting you in being what you desire to be. If you are consciously experiencing compassion are you not being compassionate?

The understanding gained from the experience develops your potential, raises your level of consciousness, and allows you to improve the quality of your life.

Discernment

Does experience alone provide understanding? Clue: Can anything in life exist by itself, or does it exist because of its relationship to something else? The twenty life experiences in Part II point out that discernment is an essential factor in understanding yourself, others, and life. The ability to distinguish obvious and subtle differences between thought, the body's senses, emotion, and consciousness is discernment. The same applies to distinguishing the obvious and subtle differences between your behavior and your personality.

Discernment adds meaning, purpose, and value to what you experience. Your perception is a good example of that. How you view everything defines and determines your behavior. And your behavior conceals your true identity from yourself and others. Discernment allows you to distinguish your personality from your behavior: your personality is your true identity. It is the personality that distinguishes one individual from another. Presently, you and those who think they know you recognize you by your physical characteristics and behavior. But your personality has no physical characteristics or behavioral traits because it is non-physical. Thus your personality is neither something that is learned nor something that is acquired: it is as unique and identifiable as the body's DNA. Once you develop your ability to recognize your personality, you realize that it remains the same from one incarnation to the next. This self-evident fact can be seen in the twenty life experiences presented. It didn't matter if their vehicle was a celebrity, a "nobody," rich, or poor. What mattered was how the personality perceived what it was experiencing, how those experiences affected the personality, and how the personality responded. These are all part of the journey and it is the journey that allows the personality to develop its potential, raise the level of its consciousness awareness, and understand itself.

Discernment is the development of an inquiring mind. An inquiring mind is free of thought. Therefore it is quiet, still and observant. It is alert, skeptical, and responsible. An inquiring mind is able to investigate because it isn't anchored to any ideology, doctrine or dogma. It is disciplined in that it never assumes or reaches a conclusion, because it understands that everything is constantly changing.

> An inquiring mind knows *how* to think
> rather than *what* to think.
>
> – *David Zimmer*

It embraces differences as opposed to sameness or more of the same. Understanding is its objective as opposed to accumulating knowledge. Such a mind is completely neutral, free of any influence, and cannot be controlled by others.

Personalities who have increased their discerning abilities are attracted to the most challenging and difficult life experiences. These personalities will have numerous incarnations that remain a part of history because they have been tested and tempered by eons of incarnations.

The Source of Resistance

What develops the personality's potential? Yes, resistance develops potential, but what provides the resistance needed to develop the personality's potential? What about thought, knowledge, images, and labels? Thought is the language of knowledge and imagery. Because thought is anchored to knowledge and imagery it unknowingly makes itself an authority. Therefore, thought is going to reject anything that conflicts with what is contained within the brain's memory and blindly accept anything that is in agreement within the brain's memory.

Thought is also fragmented, which makes it very unreliable. Magicians rely upon thought's fragmentation to perform their illusions – they realize that thought is unable to simultaneously comprehend what the magician's right and left hands are doing. The fragmentation of thought causes you to see angels as being different from humans. In turn, it creates images of angels that portray how angels are different from man. Fragmentation also prevents thought from being aware of the images it creates, and causes thought to perceive them as being real and accurate – a fact that can easily be demonstrated by observing how thought responds to anything it perceives as being different. It creates images that represent those differences, and your perception of those images stimulates an emotional response within you.

The brain combines image and emotion, transforming it into a label, thereby further transforming "what is" into "what is not". Labels impersonalize what is and cause you to do things that you normally wouldn't do. You can easily see this by observing how labels affect your behavior. You behave one way when you see yourself as a man or woman, you behave differently when you see yourself as a son or daughter, and you behave differently when you see yourself as a father or mother. Likewise you respond differently to strangers than you do to acquaintances, and to friends differently than to acquaintances or strangers.

Some labels are more impersonal and emotional than others. The more impersonal and emotional they are, the more people criticize, chastise, attack, and even kill those who appear to be different to themselves. The label "heathen" allowed the white and black people in the United States to kill thousands of American Indians. The label "witch" allowed Christians to burn and drown those who appeared to be different. Titles are labels that cause people to feel superior to others and treat others as being inferior. Labels prevent people from seeing that they are destroying their environment – the very thing their survival is dependent upon. The latter clearly illustrates that labels, as well as knowledge and images, prevent you from obtaining what you desire.

Having said that, let's return to our original question: What develops the personality's potential? Could the answer be "illusion"? Does reality conflict with or embrace illusion? Does illusion resist or embrace reality? Isn't conflict a form of resistance?

In order for the personality to understand itself it has to experience what it is not. It does this so well that it perceives that it is what is not, which is an illusion. Words and labels are illusions. They merely represent "what is." The word "steak" is merely an image that represents a type of meat. The word "steak" neither satisfies your palate nor nourishes your body. All this implies that both the physical and non-physical are worlds of illusion. If they were, they would provide an endless supply of resistance for the personality to develop its unlimited potential. And the personality would have to be very discerning to distinguish the difference between the "illusion of life" and the "reality of life".

Power

Life has constantly shown man that he has no power or control over Nature. Life has power over man because it created man and is responsible for its creation. Unfortunately, mortals don't realize that they only have power over what they create, because they are responsible for their creations. Observing how mortals attempt to control what they didn't create confirms that fact. It's also substantiated by their unwillingness to assume responsibility for what they create. Thus mankind doesn't realize that the more they think they have control, the less control they actually have.

Humankind uses force in an attempt to obtain power, because they don't understand that power arises from understanding and meaning.

> Life and everything within it is
> governed by power, not force. Cause
> and intent determine power.
>
> – *David Zimmer*

This is why power has to do with purpose and value. Force is always associated with thought, whereas power is always associated with consciousness. Power is uplifting, gracious, and noble and, unlike force, it requires no justification. Force always creates counter-force. Power, on the other hand, is always peaceful, still and silent. Force is incomplete and always consumes energy; therefore it always has to be fed. Power is whole, complete and always supports and provides energy. It makes no demands; it has no needs. Force is concrete, literal and arguable. It requires proof and support. Power requires no proof or support.

Man could see this if he asked himself, "Do my actions or behavior indicate that I am using power or force?" You can see this by asking yourself, "Do my actions indicate that I am using power or force?"

Thought

For most people, thought is so much a part of their identity that they are oblivious to its existence. I used to define thought as being the voice in your head or mind until I realized that people couldn't comprehend or relate to what I was saying. They were unable to see thought as being separate from themselves. A few even thought I was talking about hearing non-existent voices like the ones that crazy people hear.

Thought is the busy chatting or dialogue that goes on in your mind. It says such things as: "I want," "I don't want," "I need", "They should", "They shouldn't", and "I'm angry". Thought is the mind's never-ending storyteller. If you pay attention to these stories you'll see that they are always concerned about the past or future, never the present moment. Worry and fear are all products of thought. "Am I going to get the job?" "Will my spouse find out?" "Things will be better when ...", and "Am I going to get married?" Thought frequently chews on some minor issue and transforms it into a bigger one. If left unchecked, thought makes the issue so monumental that you become incapacitated by it.

Animals don't think. That's because thought is a product of language and, like primitive man, animals rely upon body language to communicate. Body language isn't actually a language in that it doesn't have syntax. Anyone who has observed birds soon realizes that their disputes don't last long and usually aren't fatal. If birds thought, they would keep the dispute alive by chewing on it (constantly thinking about it). Its story would be something like the following. "I don't believe that bird entered my territory! Who does he think he is? Talk about being inconsiderate! Boy I am going to teach him a thing or two. I wish I had kicked his butt. What if I couldn't and he ended up kicking mine? Boy, am I lucky I didn't let him get the best of me." The story would probably go on and on for hours, maybe days.

Meditation

The primary obstacle to humanity's development is their reliance upon thought and unawareness of consciousness. The absence of consciousness prevents humanity from having a reliable way to observe and interpret their environment. It prevents them from having a viable means to challenge thought and perception, and so humanity is unaware that what they call "reality" is actually the "illusion of life". **Simply put, meditation is a way to approach and experience consciousness.** And experiencing consciousness is the only way man can be aware of consciousness and observe how it is different from thought.

Many people see meditation as a separate act, like sitting in the lotus position or chanting, and they may not realize that their perception of what meditation is prevents them from being able to meditate. Others are so consumed by thought that they need the assistance of biofeedback equipment to learn how to relax and stay focused enough to be able to meditate. Most people can learn to meditate by sitting in a comfortable position and focusing their total attention on their breathing. If they practice doing this for about a week, they can actually observe thought ending. When that happens, the mind is still, quiet, and at peace.

Telling yourself not to think is thinking! You might not realize it but thought will come to an end of its own accord if you attentively observe it. In other words, one hundred percent of your attention is focused on thought. Not being able to do that is a good indication that you may need the assistance of a biofeedback system. An effective and reliable system can be purchased quite inexpensively.

There are so many different perceptions about meditating that you might have been doing it all along without realizing it. The way you meditate may not be seen as meditation because not everyone approaches or enters into meditation the same way. Some enter meditation through gardening, others through their music and others through their craft. Almost anything can be used to enter meditation; the act that opens the door to consciousness is irrelevant. What matters is the quality of mind that you bring to the act. Experiencing that state of mind allows you to understand it. Once you understand why your approach is meditative, you can transform all your acts into meditation. When you do that, it does not matter where you are or what you're doing: everything you do is meditative.

Both activity and the absence of activity (rest) are a natural function of life; both are needed to provide balance. Just as your body needs a rest from physical activity, so your mind needs a rest from the activity of thought. Meditation is the absence of thought. It is an attentive, quiet and peaceful state of being.

Meditation will appear foreign or strange to you if you predominately rely upon thought rather than your feelings, senses, emotions, and intuition. This causes you to think that you need a teacher, an instructor or a guru to teach you how to meditate. If that sounds like you, you may want to look upon meditation as an inward and outward movement like breathing: The inward movement is observing how everything affects your thoughts, feelings and emotions, and the outward movement is observing how you respond to those thoughts, feelings and emotions.

Like breathing, meditation allows you to process what you mentally "upload" on the inward movement and "download" the waste created by this process on the outward movement. It is a rejuvenating process that keeps you fresh. Just as the digestive process allows the body to transform food into energy and discard the waste, meditation allows you to transform what you mentally consume into understanding and to discard the waste. Although you probably realize the importance of paying attention to what you eat and drink, you may not be able to see the importance of paying attention to what you mentally consume. You realize the importance of exercising the body, yet you may not see the importance of exercising the mind. Similarly, you realize the importance of discarding body waste, yet perhaps you don't see the importance of discarding mental waste. Consequently you may not see the purpose nor importance of meditation.

Meditation is life's immutable and indestructible paradigm. It is immutable because consciousness prevents itself from being corrupted by constantly challenging itself and its findings in many different ways. Unlike thought, belief or perception, consciousness simultaneously examines everything from many different angles or viewing points. This omnipotent perspective provides a clearer and more accurate view. In turn, your response is more precise and effective. On the other hand, thought, belief and perception are not immutable. They can be altered, corrupted, lost and destroyed. Meditation is your personalized, unrestricted and incorruptible paradigm for living – a paradigm that is free of all limitations, dependencies, or boundaries. A paradigm that is not contingent upon chance, accident or divine intervention. A paradigm that is not dependent upon education, social position or race.

Meditation enables you to harmoniously coexist with your environment, even while living in the throes of chaos, prejudice and violence. Absolutely nothing can prevent meditation from improving the quality of your life. It doesn't matter if you are incarcerated, married to the wrong person, or living in poverty. Not even your friends, relatives and business associates can prevent you from experiencing the good life. It is the only paradigm that preserves your freedom, maintains your dignity, and restores your balance. On the other hand, the absence of meditation causes you to:

- Pay attention to thought and ignore or deny your feelings, emotions and intuition.

- Pursue, consume and store knowledge rather than question, challenge and process it into understanding.

- Value and defend your perception rather than challenge it.

- Experience the same thing over and over again.

- Rely upon so-called experts and authorities rather than yourself.

- Experience fear, conflict and sorrow rather than unconditional love, peace and joy.

Pendulum Dowsing

Dowsing is a dynamic way to observe the interaction between what cannot be seen (consciousness) and what can be seen (the physical). Pendulum dowsing is one of many different ways to dowse. Unexplainable "goose bumps," odd sensations at the nape of your neck, and cold chills down your spine are examples of how the physical body can dowse.

You can use anything as a weight (bob) that you can hang on a string, thread, or chain. It can be any size, even as small as a paper clip on a thread. The weight or bob can be made out of metal, glass, or plastic. The chain or string is usually about 3 to 4 inches long. Hold it as shown.

Start by sitting in a comfortable chair, with your feet flat on the floor. Your mind and body should feel relaxed and stress-free.

Take your pendulum and hold the string or chain between your thumb and first finger. Hold it with about 1/2 to 3 inches string length. The string length will determine how fast it will swing. Feel free to adjust the length of the string until you get the desired swing. Next, hold the pendulum out about a foot from your solar plexus so it can swing freely. You can hold you arm in a horizontal position or you can rest your elbow on a table in front of you.

Pendulum dowsing is used to ask questions that can be answered "yes" or "no". Reading a pendulum varies from person to person: the answer is determined by how the bob swings. The bob will swing:

- From side to side (right to left or left to right)

- From front to back (away from and towards your body)

- In circles (clockwise or counterclockwise)

- In an elliptical motion

Now it is time to determine which movement is a "yes" and which is a "no" for you. You may find that forward and back from the body is a yes and from side to side is a no. Or your pendulum may prefer to use a circular motion. Hold the pendulum in one hand and use the other to stop its motion. Keep your eyes focused on the bob. Once the bob is still, address the pendulum and say, "Show me a yes." The bob will soon swing in one direction. After the bob has swung a few times, say, "Show me a no." The pendulum will soon swing in the opposite direction.

Be sure the pendulum is not being guided or moved by your hand or fingers. You're ready to begin asking the pendulum questions once you have determined what a "yes" and "no" movement is. **Always get permission to use the pendulum.** Thus your first question will always be something similar to, "May I use the pendulum now to obtain assistance?"

Keep in mind that sometimes the answer is unknown for various reasons. When that happens the pendulum will either swing diagonally or in an elliptical motion. Sometimes the pendulum will appear to bob or dance up and down. If the pendulum does not move at all, stop what you're doing and try again later.

The best way to learn to read the pendulum's movements is by asking questions that you know the answers to. Then move on to questions that you don't know, but can quickly find out. Start with simple questions such as, "Is it sunny today?" "Is it Tuesday?" or "Can you answer my questions now?" Once you're comfortable using the pendulum, you can start to ask it personal questions.

The usefulness and reliability of the answers obtained by dowsing are determined by the purpose and quality of the questions you ask. It is never wise to dowse for answers when you are emotionally upset, disturbed or stressed out. It is also important to realize that everything in life is on a "need to know" basis. The pendulum will not respond if life, in its infinite wisdom, determines that you don't need to know the answer to the question you asked.

The reason for asking the question has to be for the highest good, such as seeking assistance in comprehending a concept, confirming your intuition, or discerning the validity of something. For example, you might want to ask the pendulum if there are any angels physically incarnate today. Keep in mind that it may not respond if you ask whether so and so is an angel because everything in life is on a need to know basis.

The response obtained from pendulum dowsing will always be misleading or unreliable if you attempt to use dowsing to obtain advice, lottery numbers, information about the future, or personal information about another person. The same applies to using dowsing for illegal purposes. As pointed out earlier, there is a reason for everything you experience. It's your responsibility to determine that reason; life will never assume that responsibility.

The quality of your answers obtained by dowsing is determined by how precise or specific your questions are. Consequently dowsing is a good exercise in discernment. Is asking, "Is the ring I am wearing made out of gold?" a precise question? What if you are wearing more than one ring? Does a yes answer tell you whether the ring is gold plated, gold filled, 10 ct gold, or 18 ct gold? Not everyone has the same perception of what a gold ring is.

Neither Believe Nor Disbelieve

Basically the world is comprised of disbelievers and believers. Disbelievers function on the premise that everything is false until it is proven to be true, but don't question or examine what determines whether something is true or false. Believers function on the premise that everything is true until proven otherwise, but don't question or examine what makes that determination. Both are flawed because they are dependent upon an "authority" which is a creation of thought. Truth is relative and never absolute. The higher the level of your conscious awareness, the more accurate the truth. This is why seasoned personalities neither believe nor disbelieve anything. They are always in a state of not knowing. They remain in that state by questioning, testing, and challenging everything. Unlike man, they understand that "what is" remains what it *is* regardless of how it is perceived, challenged, or tested, whereas illusion or "what is not" changes or ceases to exist.

Angels are personalities who are aware of consciousness and use its power, wisdom, and support to overcome and defeat its formidable adversaries: thought and illusion. Unlike resident personalities, angels devote twenty-one or more incarnations to examining a single concept such as responsibility, compassion, or nurture. Their experience as a resident personality allowed them to realize that understanding involves experiencing something in a wide variety of ways from many different perspectives. Angels welcome everything that life has to offer and use it to develop their potential. They realize that everything they experience serves a purpose, has potential, and has value. Angels allow life to direct them as opposed to trying to control life and others.

Angels understand that life is loving and non-judgmental. They realize that life doesn't care who you are in respect to title, nationality, race, or color. It isn't concerned about what sex you are or what your sexual orientation is. Life isn't concerned about what your religion is, which side of the law you are on, or what your occupation is. Your perception or view of life doesn't alter or change what life is or does. Life treats everyone in the same just, fair, and impartial way. Life is completely non-judgmental, even if you curse it. Like life, angels are only interested in the journey for it is the never-ending journey that allows them to develop their unlimited potential, raise their level of consciousness, and understand themselves.

Summary

Examining what you have in common with angels enabled you to see that angels basically experience the same things you do. Thus you cannot recognize angels by what they do. If every personality uses the human body as a vehicle then you cannot recognize an angel by his or her appearance. Eventually your examination of angels causes some aspect of you to ask, "What is an angel?"

What makes that question insightful? Is that which asked the question acknowledging that it doesn't know what an angel is? Would thought ask that question? Why? If thought didn't ask it, what did? The question, as well as your inability to answer it, implies that there is an aspect of you that neither thought nor you are aware of.

Only that which is whole and complete is able to see and acknowledge what it doesn't understand. It never assumes. Either it understands something or it doesn't.

Knowing and understanding are two completely different things. Understanding is the negation of knowledge through experience. Consciousness is always in a state of not knowing, even when it understands something. Unlike thought, consciousness always acknowledges what it doesn't understand.

Debating, arguing or defending your perception creates conflict and prevents you from raising your consciousness. The same is true when you reject, deny or avoid anything that conflicts with your perception. Another source of conflict is thinking, believing or deducing that you "understand" what you know. Whenever you do any of these your query or investigation ends because the mind does not see any sense in examining what it knows. When you acknowledge that you do not know something, you instruct the mind to observe and investigate. It is impossible to observe or investigate anything when thought is rehashing the past or chewing on what may occur in the future. That is why it is so important to set aside at least twenty minutes a day to allow the mind to quiet down. Thought will come to an end if you attentively observe it. Quiet, peaceful and attentive observation is meditation – those who meditate understand this from experience. They realize the importance of scheduling some quiet time at the same time each morning or evening and in the same location.

* * *

We hope you found something of interest in *Living Among Angels*. It's the first of three books by us. The other two books in the trilogy are *Living Among Aliens* and *Living Among the Trinity*. Each book builds upon the previous one and helps you obtain a better understanding of life. The more you understand about life, the more you realize that it has no secrets: life is completely transparent, making it impossible for it to have any secrets. Everything about life is in plain sight of all for those who have eyes that know how to see. Your unexamined images and perception block your sight and prevent you from experiencing a reality free of secrets. A reality free of secrets can be very disturbing if you're trying to hide something or have secrets. You also have to consider the possibility that you have some secrets that you have unknowingly hidden from yourself! Secrets cause you to play perplexing and terrifying games that encapsulate your freedom and stagnate your life.

Addenda

Addendum I: Saint Peter and Aaron Smith

An illuminating question always stimulates additional questions rather than answers. Understanding is always the motivation for asking illuminating questions; it is never about obtaining more knowledge.

Aaron Smith was like the majority of people who had some religious instruction when he was younger. He believed that he would go to heaven when he died because he was basically a good man. He also perceived that his experience in heaven would be better than what was experienced on earth. On the eve of Aaron's 39th wedding anniversary he had an unexpected heart attack and died. The following story about Aaron is very similar to millions of others.

"Welcome, Aaron Smith, we rejoice at your return," said St. Peter.

With a confused look on his face, Aaron stammered, "Where am I? What happened? I mean, one moment I was surrounded by my family, my mom and dad, my grandparents, my older brother, they were all there. They all looked just as I remembered them. Then suddenly they're gone and I'm here with you! Who are you and what did you do with my family?"

"Of course they are as you remember them. Before you were seeing them with your eyes but here you see them through your heart and all the love it holds for them. They all wanted to be present at your arrival: they love you and are overjoyed to see you again. You will rejoin them momentarily but before doing so it is necessary for you and me to talk. I am St. Peter, Aaron, and this is your new home. What you call the hereafter, glory land, or heaven. You do remember what happened, do you not?"

Suddenly Aaron recalled his last moments on Earth and, although it was as if he were reliving the experience, there was a complete absence of pain. Instead he became aware of the enveloping warmth and peace that suddenly overtook him as he was freed from the confines of his earthly vehicle. Now here he was standing in front of St. Peter.

Looking around, Aaron asked, "If you are St. Peter, where are the Pearly Gates?"

"You already passed through what is called the Pearly Gates, known to some as the Golden Gate. This is merely the name that has been given to the passage from the physical realm to the realm of the spirit," St. Peter replied. "Is there anything else you wish to ask?"

Here he was. Standing in front of St. Peter, the St. Peter that he had heard of and read of throughout his life. This was his opportunity to ask any question, any at all, but all he could think of was how gently the words of this being came to him. It reminded him of how filled he was with love when first he held one of his newborn children, and it felt as though it certainly must spill over because it was too much to be contained within him.

"Aaron, do you have any more questions?"

"You are real! I'm so happy to be here with you, I mean, to finally meet you," he stammered. "I, ah, I guess I was expecting something different. I remember something about looking at my Book of Life but I'm sorry to say I never really quite understood what that meant. Is that what we need to talk about – are you going to show me that? I bet that's it, right?"

Smiling, St. Peter replied, "The Book of Life. We don't get too many arrivals who do understand the meaning of those words, Aaron. It is referencing your review of the most recent chapter of your life. You may complete the review at your leisure, however and whenever you feel you are ready."

"However? What do you mean?"

"Some choose to have guidance as they go through their review, to have some other spirit there with them. Others feel more comfortable doing it alone; it is all up to you. And I can assure you that you will come to that moment when you are aware that you must examine the experiences of Aaron Smith. Then will be the time, then you shall do so. Do you understand?"

"I think so. I mean, ah, it kind of makes sense to do it that way, I guess, at least to me." Aaron answered, becoming aware that he was stammering once again as he searched for an intelligent reply.

"Hey, just a minute here! If I'm not here with you to take a look at my life, then just exactly what am I doing here?" He felt he had a right to demand an answer to that question – after all, his reunion with his family had been rather rudely interrupted.

"Very good! Yes, we must turn to the reason you are here with me instead of with those you love."

St. Peter's response to his outburst embarrassed Aaron. Again he found himself stammering for words. "I, ah, I'm sorry, sir. I didn't mean to make it sound as though I don't want to be here with you. I mean, I guess I'm not sure what I mean. I just want to do whatever it is that I have to do and I don't know what that is."

"There is nothing that you have to do right now, Aaron. That is not the purpose of our meeting. We are here, in the presence of one another, that I may present to you a most important question. It will not require that you provide an immediate answer; instead, it is preferred that you ponder the question from time to time until the answer appears to you in your heart."

St. Peter now appeared most solemn and Aaron found himself feeling apprehensive, wondering what the question could be.

"What do you want to do for the rest of eternity, Aaron? That is the question that we ask you to examine until you come to discover your answer."

"I don't understand," Aaron blurted out. "I thought this was heaven! I thought I would happily spend the rest of eternity here with my loved ones."

"You are about to join the others. But, Aaron, eternity? That is a long, long time. You must remember how active your earthly life was. Why, you were always doing something, whether it was fishing, working on your house or busy with your employer. You spent many hours with your grandchildren as well as with friends, did you not?"

"Yes, that is so. I had to keep busy doing something. I didn't like to be idle."

"Why not? Why didn't you like to be idle," St. Peter queried.

Aaron's answer was simply, "I got bored when I had nothing to do."

"And so you would find something to do, isn't that correct?"

"Well, yes. As long as I can remember I had to be busy doing something. Even when I was too tired to do something I would watch television, or read a book or something."

"And now that you are here in heaven, you believe it will be otherwise? You believe you will be happy just doing nothing? Let me remind you once again that we are talking about eternity and that means for ever and ever. Do you believe you can escape boredom while doing nothing simply because you now discover yourself to be in heaven? Aaron, I'm not asking this question to cause you any anxiety whatsoever. Remember, you need not answer it now. Indeed, you are not expected to answer it when first asked. You will find moments when the question returns to your thoughts, moments when you shall then begin to ponder this question. There is no immediacy placed upon the discovery of your answer. After all, you have eternity to do so if you so choose."

"Let me see if I understand what you are telling me. The question is, 'what do I want to do for the rest of eternity'. I don't have to answer you now and I'm really glad to hear that because I don't have the faintest idea of how I would answer. But, when we are done here, I will rejoin my family. There will be moments when I will recall the question and then I should examine myself to try to determine my response. Is that correct?"

"That is an accurate summation of our discussion," St. Peter responded.

"And then what? When I know the answer, what do I do? Do I come find you or who do I give my answer to?"

"You need not concern yourself with that now, Aaron. Simply put, when you know your answer to the question you will also be aware of what to do next. All you need to do now is relax and enjoy this moment. Return to those who await you and bask in their love. Someday, you and I shall meet again. Until then know that it has been my privilege to serve you."

With those parting word St. Peter disappeared. All those who had preceded him in death again surrounded Aaron. It was most joyous.

St. Peter had moved on to the next arrival. Aaron would eventually discover his answer, he knew. They all did. There would come that moment when they realized there was something more that they wanted to experience, something more they wanted to understand. Then they would be gone, off on another adventure. St. Peter smiled as he thought of how wondrous and beautiful it was. "It is all about the journey," he said to himself, "Yes, life is just about the never-ending journey."

No, I didn't eavesdrop on one of St. Peter's orientation interviews. This story effectively points out how you can use questions to challenge your beliefs. I posed the question, "What do you want to do for the rest of eternity?" to myself when I started examining my concept of heaven a long time ago. The question caused me to pause and ponder the idea of spending eternity in heaven. Eternity is indeed a long, long time.

Yes, you do have moments here when you are bored. Is it possible that the reason that you're *not* bored at times is because you're "doing" something; participating in some activity? What are your expectations for your afterlife? How do you perceive you will spend the rest of eternity once you leave this plane of existence? In other words, "What do you want to do for the rest of eternity?"

Note: Of course, St. Peter is not really waiting at the pearly gates for us when we die. I've merely used this fictional idea to illustrate the story.

Addendum II

Recommended Reading

Additional insights into thought and how it affects people can be obtained by doing the work described in *Loving What Is* by Byron Katie. When you have done that you might want to examine her latest book, *I Need Your Love – Is That True?*

I also suggest *A New Earth: Awakening to Your Life's Purpose* by Eckhart Tolle. He also wrote a New York Times best seller, *The Power of Now*.

Juddu Krisnamurti was probably one of the most influential spiritual teachers of the twentieth century. Although he has written several books, we recommend *Freedom from the Known and Awakening of Intelligence*.

If you prefer lighter reading, I recommend *Mister God, This Is Anna* by Fynn. It's a delightful book that illustrates the power of curiosity, questioning, and examining things from different perspectives. In many ways Anna reminds me of Inquisitive Nattie in Chapter Five.

Who Moved My Cheese? by Spencer Johnson, M.D. is a simple story about change.

Addendum III

Glossary

Akashic Records. God's memory, which is the embodiment of the atom absolute known as Intelligence or the Holy Ghost. One of three immutable, incorruptible and indestructible standards in life.

alien. A personality who first took physical form on a planet other than the one it is presently incarnated on.

angel. A seasoned personality (permanent atom) serving a tour of duty as a role model.

archangel. The embodiment of one of the seven Universal Laws. One of the seven aspects of God.

Atom Absolute. A perfect, pure, and complete personality of the Consciousness Absolute. The collective expression of the three Atoms Absolute personify the personality of God.

channeler. An individual acting as a medium or intermediary for a personality that is not incarnate.

companion personalities. Personalities that have worked closely with another in several incarnations.

consciousness. The soul's mind. A mirrored image of the Consciousness Absolute.

Consciousness Absolute. God's mind. One of three immutable, incorruptible and indestructible standards in life.

conscious awareness. A qualitative state of "being" where the personality is simultaneously aware of its physical, mental and spiritual nature.

emotion. The language of the personality.

fallen angel. An angel that is physically incarnate. Many falsely perceive that a fallen angel fell from grace or did something wrong.

feelings. The language of the human body's senses.

free will. The personality's ability to choose between acting upon thought or upon consciousness.

guardian angel. A non-incarnate personality serving as a mentor, tutor, or teacher for one or more incarnate personalities.

incarnation. One of a succession of periods spent as a human being.

Intelligence. The implementation of God's law. Also known as the "spirit" of the law. The atom absolute commonly referred to as the Holy Ghost in the Trinity.

karma. Life's response to an imbalance created by a personality's flawed perception. A cause and effect relationship pertaining to balance.

life. Life=creation+ reincarnation + evolution. The whole and everything that comprises it: Naught, the Atoms Absolute, Consciousness Absolute and their expressions.

Love. The expression of God's Law. The atom absolute commonly referred to as the Son in the Trinity. Unlike what humanity perceives or calls love, this love is unconditional.

naught. The first and only "real" plane of existence. The source of everything.

perception. Unexamined knowledge, experiences, and beliefs that color, alter, and distort the personality's vision.

permanent atom. Individual personalities that collectively express the consciousness of mankind which is the soul. See **personality**. Every permanent atom is endowed with Will, Love, and Intelligence from the three atoms absolute.

personality. A permanent atom's or atom absolute's true identity. The identifying character traits of the permanent atom. Commonly and falsely known as the "soul".

personality group. A group of personalities that have worked together in previous incarnations. Falsely known as "soul group".

reincarnation. One of three primary processes of life. See **life**.

resident personality. An incarnating personality serving its first tour of duty. The planet it first incarnates on is its planet of origin, its residence.

review incarnation. The last incarnation in a personality's tour of duty. Its purpose is to assist the personality in reviewing and correcting any misconceptions it created during its tour of duty, as well as to allow the personality to realize what it accomplished.

soul. The consciousness of the permanent atoms. The mirrored expression of the Consciousness Absolute.

thought. The language of perception. A mirrored image of the **soul**. Unlike the soul it is fragmented, incomplete and binary.

Universal Laws. The laws that govern all life. One of three immutable, incorruptible and indestructible standards in life.

Will. The law of God and creative force of everything. The atom absolute that is commonly referred to as "the Father" in the Trinity.

David Zimmer

When I look back on my life, I see what an amazing journey it has been. What makes it so amazing is how messed up I was physically, mentally, and emotionally when I began my journey thirty years ago, at the age of 34. If that wasn't enough, I had to overcome various limitations such as depression and a type of dyslexia that affects both my physical eyesight as well as my hearing. I went from hardly being able to read and verbally murdering the English language to reading the Akashic Records at will and doing past life seminars. I went from being unable to see or understand simple concepts to understanding life's complex relationships and the immutable laws that govern life. I went from being oblivious to my surroundings to being extremely consciously aware. I ended the hell I had unknowingly created, and the quality of my life has improved far beyond anything I could have possibly imagined. In fact, it keeps on getting better every week, year after year!

As I raised my conscious awareness over the past thirty years, new abilities and opportunities to express them unfolded. My investigation of reincarnation, creation, and evolution enabled me to do things I'd thought were impossible. One of them was the ability to read the Akashic Records. That experience motivated me to do seminars on reincarnation and past life readings. Examining thousands of past lives and my involvement with angels that are incarnate today enabled me to realize that you have to rely upon consciousness rather than thought or knowledge to destroy the illusions created by thought. Everyone has to discover his or her unique path and consciously travel it. No one else can take that path and do your work.

Luella Stroh

I have been examining creation, evolution, and reincarnation for more than 25 years. These examinations enabled me to raise my conscious awareness and achieve a "quiet mind". About two years after reaching this state of being, I discovered that I could read the Akashic Records, and I used that ability to further my understanding of creation, evolution, and reincarnation. After several unexpected events, I found myself working with angels and other spiritual entities living amongst us for the next twelve years. Eventually I realized that it was time to move on and experience something different, and I decided to finally do what so many have encouraged me to do: publish my experiences so that others might benefit from them.

For some unexplained reason, I really don't have an image of myself. On occasion, my retired body reminds me that I'm not as young as I used to be. The following poem, which I wrote a while back, probably best represents me.

Who Am I? You Ask

You look and think you see
a middle-aged hippie,
left over from the sixties.

Ah, but in the sixties
I was baking chocolate chip cookies,
going to PTA meetings,
worrying about taxes, and
I wallpapered the kitchen.

"What's that? What happened?" you ask
I grew weary of the many
suits I was wearing.
I am taking them off,
one by one.
Soon, I can walk naked in the world.

So, what about the kids?
Oh, they're baking chocolate chip cookies,
going to PTA meetings,
worrying about taxes, and
this year they're going
to wallpaper the kitchen.

Other Books in the Series

This book is the first in the *Living Among* trilogy. The other two are:

Living Among Aliens

A scientific and philosophical examination of aliens and humankind's perception of them. The book also features actual life experiences of seven aliens living among us.

Living Among the Trinity

A self-evident examination of the Trinity and the Universal Laws that govern life. Understanding these laws allows you to effectively change the quality of your life.

For further information about these books, please see our web site:

www.trackerpress.com

We are always happy to hear your feedback about our books. Please use the following email addresses to contact us:

David: david@trackerpress.com

Luella: luella@trackerpress.com

For general **editorial enquiries and comments**:

editor@trackerpress.com

To be added to our **mailing list**:

mailing-list@trackerpress.com

Printed in the United States
67779LVS00001B/22-69